Basic Maths Solutions for CSEC

Christian A. Hume

ISBN 978-976-8260-88-8

CONTENTS

Preface

This is not a textbook. This is a book of worked examination questions. In this book you will not find detailed explanations of mathematical topics; you will not find exercises to practice mathematical topics. In this book you will find worked examination questions from the June 2002 to June 2008 CSEC Basic Proficiency Examination in Mathematics.

Even though the Basic Proficiency has been discontinued, the past papers from past 'Basic' exams are a rich and fertile resource that can provide an invaluable introduction to examination type questions for students in Form 2 to Form 4; while at the same time providing an excellent platform for the Form 5 student to pivot to the more difficult questions and topics of the General Proficiency Examination.

The solutions are presented in my own handwriting - calligraphically flawed though it may be for two reasons. The first reason is that I wish to bring a little spark of human warmth and connection to what can sometimes be a cold and unfeeling world of mathematics texts. The second and more important reason is to present the step-by-step approach to solving exam questions while making ample use of the space provided on a typical page on which an exam is written. The neatness and step by step progression of your written exam responses make a huge difference, and proper space utilization leaves sufficient room for corrections to be made and inserted if and when they become necessary.

As is the case so often in mathematics, some of the solutions presented herein represent just one of a handful of ways to solve the particular problem. Where you become aware of an alternative path to the solution to a particular problem, I urge you to follow your curiosity – explore alternative methods and then choose the one you are most comfortable with.

The examinations whose solutions are presented in this book can be downloaded for free online at: https://payhip.com/b/GuqH

Christian A. Hume

June 2002

1 (a) $366 \div 0.0012 =$

(i) 305,000

(ii) 3.05×10^5

(b) $1\frac{1}{6} \div \left(1\frac{2}{3} + 1\frac{1}{4}\right)$

$\Rightarrow \frac{7}{6} \div \left(\frac{5}{3} + \frac{5}{4}\right)$

$\Rightarrow \frac{7}{6} \div \left(\frac{(4)(5) + (3)(5)}{(3)(4)}\right)$

$\Rightarrow \frac{7}{6} \div \left(\frac{20 + 15}{12}\right)$

$\Rightarrow \frac{7}{6} \div \frac{35}{12} \qquad \Rightarrow \frac{7}{6} \times \frac{12}{35} \qquad \frac{7}{6} \times \frac{12^2}{35_5}$

$= \frac{2}{5}$

(c) Ratio of men to women at beach = 6:7

There are 28 women
Let number of men = x

$\therefore$ $x : 28 = 6 : 7$

$\Rightarrow$ $\dfrac{x}{6} = \dfrac{28}{7}$

$\Rightarrow$ $\dfrac{x}{6} = 4$

$\Rightarrow$ $x = 6 \times 4$
$ = 24$

$\therefore$ There are 24 men

$\therefore$ Number of people on the beach
$= 24 + 28$
$= 52$

2 (a) Profit = Selling Price − Cost Price
$$= \$24,400 - \$20,000$$
$$= \$4,400$$

% Profit $= \dfrac{Profit}{Cost\ Price} \times 100$

$$= \dfrac{4400}{20000} \times \dfrac{100}{1}$$

$$= 22\%$$

(b) (i) Cost of advertisement of 12 words

$$= \$60.00$$

(ii) Cost of advertisement of 24 words

$$= \$60.00 + [\$3.00 \times (24 - 16)]$$
$$= \$60.00 + (\$3.00 \times 8)$$
$$= \$60.00 + \$24.00 \qquad = \$84$$

(iii) Cost of advertisement = \$198

Cost for words in excess of 16 = \$198 - \$60

$$= \$138$$

$\therefore$ No. of words in excess of 16 $= \dfrac{138}{3}$

$$= 46$$

$\therefore$ Total number of words in advertisement

$$= 16 + 46$$
$$= 62$$

3 (a) $m = 3 \; ; \; n = -2$

(i) $mn^2 = (3)(-2)^2$

$= 3(4)$

$= 12$

(ii) $(m+n)(m-n) = m^2 - n^2$

$= (3)^2 - (-2)^2$

$= 9 - 4$

$= 5$

(b) $\dfrac{x-3}{3} + \dfrac{x+2}{4} \Rightarrow \dfrac{4(x-3) + 3(x+2)}{(3)(4)}$

$= \dfrac{4x - 12 + 3x + 6}{12}$

$= \dfrac{4x + 3x - 12 + 6}{12}$

$= \dfrac{7x - 6}{12}$

(c) $\quad 1 + 3(x-1) = 4$

$\Rightarrow \quad 1 + 3x - 3 = 4$

$\Rightarrow \quad 1 - 3 + 3x = 4$

$\Rightarrow \quad -2 + 3x = 4$

$\Rightarrow \qquad 3x = 4 + 2$

$\qquad\qquad = 6$

$\Rightarrow \qquad x = \dfrac{6}{3}$

$\qquad\qquad x = 2$

4. (a) $3x + y = 1$ —— ①
   $x - 2y = 5$ —— ②

① × 1: $3x + y = 1$ —— ①
② × 3: $3x - 6y = 15$ —— ③

───────────────────

① − ③: $y - (-6y) = 1 - 15$

   $y + 6y = -14$

   $7y = -14$

   $y = \dfrac{-14}{7}$

   $= -2$

Subst. $y = -2$ into eqn ①:

   $3x + (-2) = 1$

⟹ $3x - 2 = 1$

⟹ $3x = 1 + 2$

⟹ $3x = 3$

⟹ $x = 3/3$

⟹ $x = 1$ ∴ $x = 1, y = -2$

(b) (i) Perimeter of triangle $= (x+2) + (2x) + (3x-2)$

$$= x + 2 + 2x + 3x - 2$$
$$= x + 2x + 3x + 2 - 2$$
$$= 6x$$

(ii) Area of triangle $= \dfrac{(x+2)(2x)}{2}$

(iii) Perimeter $= 24\,cm$

$$\therefore \quad 6x = 24$$
$$x = \frac{24}{6}$$
$$= 4\,cm$$

(iv) Length of longest side $= 3(4) - 2$

$$= 12 - 2$$
$$= 10\,cm$$

§ (a)(i) £1 sterling = (1 × 2.94) Barbados dollars
            = $2.94 Barbados

(ii) $30.00 Barbados = $\left(\dfrac{30}{2.94}\right)$ pound sterling

            = £10.20 sterling

(iii) $300.00 US = (300 × 1.99) Barbados dollars

            = $597 Barbados

(b)(i) Basic hourly rate = $\dfrac{320}{40}$

            = $8/hr

(ii) Overtime hourly rate = 8 × 1.5
            = $12/hr

(iii) Total wage = Basic wage + Overtime wage

$$= \$320 + (10 \times \$12)$$

$$= \$320 + \$120$$

$$= \$440$$

(iv) Total wage = Basic wage + Overtime wage

$\therefore$ $\quad \$560 = \$320 +$ Overtime wage

$\therefore$ $\quad \$560 - \$320 =$ Overtime wage

$\therefore$ $\quad$ Overtime wage $= \$240$

$\therefore$ $\quad$ No. of overtime hours $= \dfrac{\text{Overtime wage}}{\text{Overtime rate}}$

$$= \frac{240}{12}$$

$$= 20 \text{ hrs}$$

6 (q) (i)

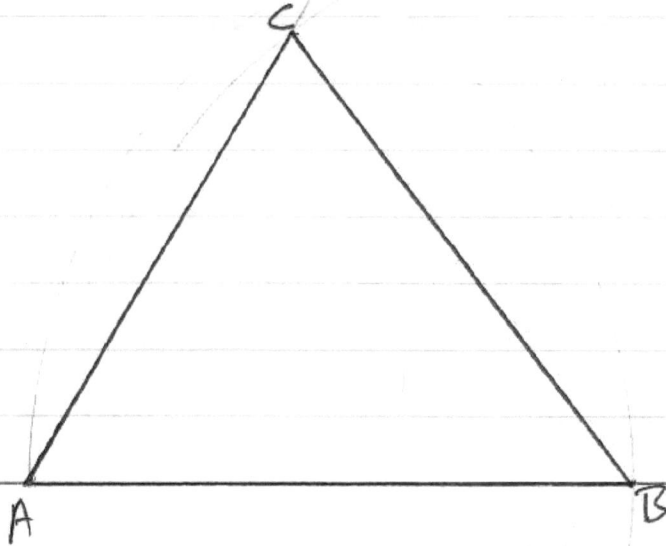

(ii) Angle $A\hat{B}C = 53°$

(b) (i) $\dfrac{QS}{20} = \sin 30$

$\Rightarrow \quad QS = 20\sin 30$

$= 20 \times 0.5$

$= 10\,cm$

(ii) $\tan x° = \dfrac{9}{10}$

$= 0.9$

$\therefore \quad x° = \tan^{-1}(0.9)$

$= 42°$

7 (a) The mode is 4

(b)(i) Mean number of children per household

$$= \frac{4+1+5+4+3+7+2+3+4+1}{10}$$

$$= \frac{34}{10}$$

$$= 3.4$$

(ii) Rearrange numbers in ascending order :-

1, 1, 2, 3, 3, 4, 4, 4, 5, 7.

The position of the median $= \frac{N+1}{2} = \frac{10+1}{2}$

$$= \frac{11}{2} = 5.5$$

∴ The mean of the 5th and the 6th numbers is the median $= \frac{3+4}{2} = \frac{7}{2}$

$$= 3.5$$

(c) The number of children in a household will always be an integer, and since the mean and median in this case are not integers, then the mode is the best average to represent the data in this survey.

(d)(i) P(household has exactly 4 children)

$$= \frac{\text{No. of households with exactly 4 children}}{\text{Total no. of households}}$$

$$= \frac{3}{10}$$

(ii) P(household has more than 4 children)

$$= \frac{\text{No. of households with more than 4 children}}{\text{Total no. of households}}$$

$$= \frac{2}{10}$$

8. (a) $A(-3,1)$; $B(-1,4)$; $C(-3,4)$

(b) $\triangle ABC$ is mapped onto $\triangle PQR$ by a translation of $\begin{pmatrix} -3 \\ -7 \end{pmatrix}$.

(d) $N(1,4)$; $M(7,4)$; $L(1,-4)$

(e) (ii) $\triangle ABC$ is mapped onto $\triangle LMN$ by an enlargement of scale factor 3 about the point $T(-5,4)$.

CARIBBEAN EXAMINATIONS COUNCIL
SECONDARY EDUCATION CERTIFICATE
EXAMINATION
MATHEMATICS
Paper 02 - Basic Proficiency

Answer sheet for Question 8.   Candidate number . . . .

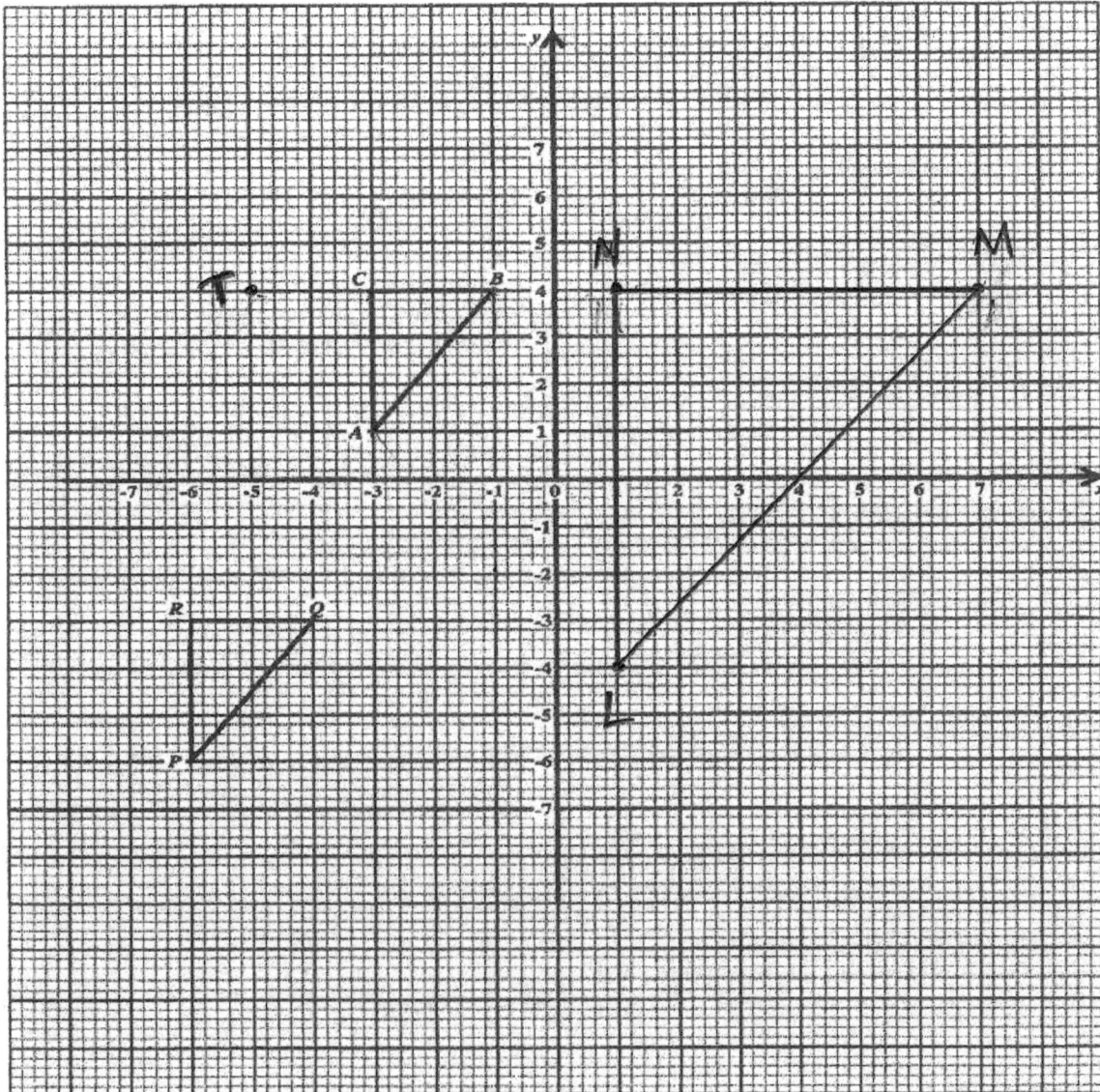

ATTACH THIS ANSWER SHEET TO YOUR ANSWER BOOKLET

000582/F 2002

9 (a) (i) Departure time = 16:40 + 35 minutes
$$= 17:15$$

(ii) Arrival time = 23:07 − 12 minutes
$$= 22:55$$

(iii) Flight time = 22:55 − 17:15
$$= 5 \text{ hrs, } 40 \text{ minutes}$$

$$= (5 \times 60) + 40$$
$$= 300 + 40$$
$$= 340 \text{ minutes}$$

(b) (i) Length of AB = $\dfrac{60}{360} \times \dfrac{2(3.14)(10)}{1}$

$$= \dfrac{1}{6} \times \dfrac{62.8}{1}$$

$$= \dfrac{62.8}{6}$$

$$= 10.47 \text{ cm}$$

(ii) Area of AOB $= \dfrac{60}{360} \times \dfrac{(3.14)(10)^2}{1}$

$$= \dfrac{1}{6} \times \dfrac{(3.14)(100)}{1}$$

$$= \dfrac{314}{6}$$

$$= 52.33 \text{ cm}^2$$

(iii)

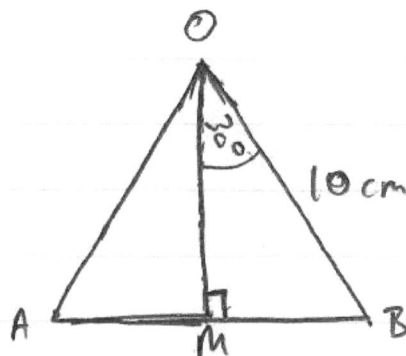

om is the perpendicular bisector of angle AOB and the chord AB.

$\therefore \dfrac{MB}{10} = \sin 30$

$\Rightarrow MB = 10 \sin 30$

$AM = MB$, $\therefore AM = 10 \sin 30$

$$\therefore AB = AM + MB$$
$$= 10 \sin 30 + 10 \sin 30$$
$$= 20 \sin 30$$
$$= 20 \left(\tfrac{1}{2}\right)$$
$$= 10 \, cm$$

10. (a) (i) gradient of $l = \dfrac{0 - (-3)}{1 - 0}$

$= \dfrac{3}{1}$

$= 3$

(ii) Equation of l: $y = 3x - 3$

(b) (i)

x	-2	-1	0	1	2	3
y	7	0	-3	-2	3	12

$y = 2x^2 - x - 3$

When $x = 0$, $y = 2(0)^2 - (0) - 3$

$= 0 - 3$

$= -3$

When $x = 1$, $y = 2(1)^2 - (1) - 3$

$= 2(1) - 1 - 3$

$= 2 - 4$

$= -2$

(iii) From graph, $2x^2 - x - 3 = 0$ for $x = -1$, and $x = 1.45$

CARIBBEAN EXAMINATIONS COUNCIL

SECONDARY EDUCATION CERTIFICATE

EXAMINATION

MATHEMATICS

Paper 02 - Basic Proficiency

Answer sheet for Question 10 (b). Candidate number ....

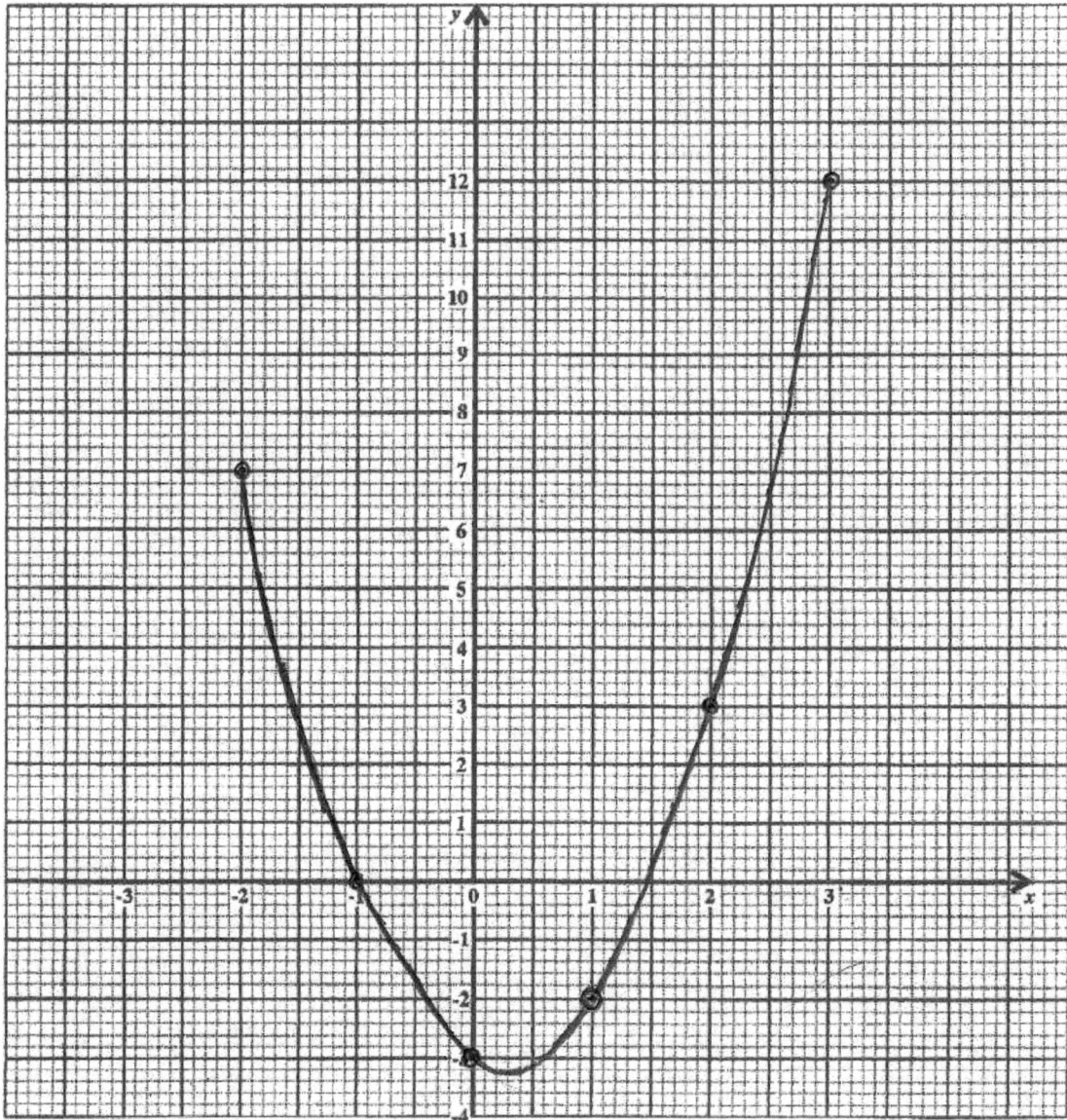

ATTACH THIS ANSWER SHEET TO YOUR ANSWER BOOKLET

000582/F 2002

June 2003

1. (a) $\dfrac{\frac{2}{3}+\frac{5}{12}}{1\frac{1}{4}}$ $\Rightarrow \dfrac{\frac{8}{12}+\frac{5}{12}}{\frac{5}{4}}$

$$\Rightarrow \dfrac{\frac{13}{12}}{\frac{5}{4}}$$

$$\Rightarrow \frac{13}{12} \times \frac{4}{5}$$

$$= \frac{13}{\underset{3}{\cancel{12}}} \times \frac{\cancel{4}}{5}$$

$$= \frac{13}{15}$$

(b) \$30 divided into ratio 2:3.

No of parts = 2+3
= 5

∴ Value of each part = $\dfrac{\$30}{5}$

= \$6

∴ Amt. of larger share = 3 × \$6

= \$18

(c) (i) Total mass of three athletes = 52.5 + 47.8 + 53.9

= 154.2 kg

(ii) Mean mass of relay team = 50.9 kg

∴ Total mass of four athletes = 50.9 × 4
= 203.6 kg

$\therefore$ Mass of the fourth athlete
$$= 203.6 - 154.2$$
$$= 49.4 \text{ kg}$$

2. (a) $\qquad$ $3x + 2 = 12 - 2x$

$\Rightarrow \quad 3x + 2x = 12 - 2$

$\Rightarrow \qquad 5x = 10$

$\Rightarrow \qquad x = \dfrac{10}{5}$

$\qquad\qquad = 2$

$\qquad \therefore \quad x = 2$

(b) $\qquad 3(5x + 2) - 2(4x + 3)$

$\Rightarrow \quad 15x + 6 - 8x - 6$

$\Rightarrow \quad 15x - 8x + 6 - 6$

$\Rightarrow \qquad 7x + 0$

$\Rightarrow \qquad 7x$

(c) $\quad f * g = f + g^2$

$\therefore \quad 5 * 2 = 5 + 2^2$
$= 5 + 4$
$= 9$

(d) $\quad 8x + 2y$

$\underline{\underline{3}}$ (a)

$$3x - y = 10 \qquad - \text{①}$$
$$2x + y = 5 \qquad - \text{②}$$

---

①+②:
$$5x = 15$$
$$x = \frac{15}{5}$$
$$= 3$$

Subst. $x=3$ into eqn ①:

$$3(3) - y = 10$$
$$9 - y = 10$$
$$9 - 10 = y$$
$$-1 = y$$

$$\therefore \quad x = 3, \quad y = -1$$

(b) (i) $x + (x+60) + (x-10) = 530$

$\Rightarrow$ $x + x + 60 + x - 10 = 530$

$\Rightarrow$ $x + x + x + 60 - 10 = 530$

$\Rightarrow$ $3x + 50 = 530$

$\Rightarrow$ $3x = 530 - 50$

$= 480$

$\therefore \quad x = \dfrac{480}{3}$

$= 160$

$x = 160$

(ii) $\therefore$ Amt. Ryan earns $= 160 + 60$

$= 220$

$\therefore$ Ryan earns \$220

4 (a) (i) Total amount customer pays $= 1.15 \times 1600$
$$= \$1,840$$

(ii) Sale price is 20% less than original price.
∴ Sale price is 80% of original price.

Let original price $= x$

∴ Sale price $= 0.8x$

∴ $\quad 0.8x = 1600$
$$x = \frac{1600}{0.8}$$

$$= \$2,000$$

(b) Total bill $= \$15 + (0.02 \times 3850)$

$$= \$15 + \$77$$

$$= \$92$$

$\underline{\underline{5}}$ (a) (i) Radius of semi-circle = 3cm

(ii) AO = 4cm

(b) Angle $A\hat{O}B$ is a right angle. Therefore triangle AOB is a right-angled triangle.

Using Pythagoras' Theorem:
$$AB^2 = AO^2 + OB^2$$
$$= 4^2 + 3^2$$
$$= 16 + 9$$
$$= 25$$

$$\therefore AB = \sqrt{25}$$

$$= 5cm$$

(c) (i) Perimeter $ABCD = AB + BCD + DA$

$$= 5 + BCD + 5$$

$$BCD = \frac{1}{2} \cdot 2\pi r \quad (r = \text{radius of semi-circle})$$

$$= \frac{1}{2} \times 2 \times 3.14 \times 3$$

$$= 9.42 \text{ cm}$$

$\therefore$ Perimeter $ABCD = 5 + 9.42 + 5$

$$= 19.42 \text{ cm}$$

(ii) Area of $ABCD$ = Area of triangle ABD
+ Area of semi-circle DBC.

Area of triangle $= \dfrac{6 \times 4}{2} = \dfrac{24}{2} = 12 \text{ cm}^2$

Area of semi-circle $= \dfrac{1}{2} \pi (3)^2 = \dfrac{9\pi}{2} = \dfrac{9 \times 3.14}{2}$

$$= \dfrac{28.26}{2}$$

$$= 14.13 \text{ cm}^2$$

$\therefore$ Area of ABCD $= 12 + 14 \cdot 13$
$$= 26 \cdot 13 \ cm^2$$

6.(a) $P(2,2)$; $P'(-2,2)$

(b) The centre of rotation is the origin

The angle of rotation is $90°$

The direction of rotation is anti-clockwise

(c) For $\triangle PQR$: $P(2,2)$; $Q(5,2)$; $R(2,6)$

Corresponding coordinates for $\triangle P''Q'''R'''$

after a translation by the vector $\begin{pmatrix} -6 \\ -6 \end{pmatrix}$:

$P''' = \begin{pmatrix} 2 \\ 2 \end{pmatrix} + \begin{pmatrix} -6 \\ -6 \end{pmatrix} = \begin{pmatrix} 2-6 \\ 2-6 \end{pmatrix} = \begin{pmatrix} -4 \\ -4 \end{pmatrix} \Rightarrow (-4,-4)$

$Q''' = \begin{pmatrix} 5 \\ 2 \end{pmatrix} + \begin{pmatrix} -6 \\ -6 \end{pmatrix} = \begin{pmatrix} 5-6 \\ 2-6 \end{pmatrix} = \begin{pmatrix} -1 \\ -4 \end{pmatrix} \Rightarrow (-1,-4)$

$R''' = \begin{pmatrix} 2 \\ 6 \end{pmatrix} + \begin{pmatrix} -6 \\ -6 \end{pmatrix} = \begin{pmatrix} 2-6 \\ 6-6 \end{pmatrix} = \begin{pmatrix} -4 \\ 0 \end{pmatrix} \Rightarrow (-4,0)$

CARIBBEAN EXAMINATIONS COUNCIL
SECONDARY EDUCATION CERTIFICATE
EXAMINATION
MATHEMATICS
Paper 02 - Basic Proficiency

Answer sheet for Question 6.

Candidate number . . . .

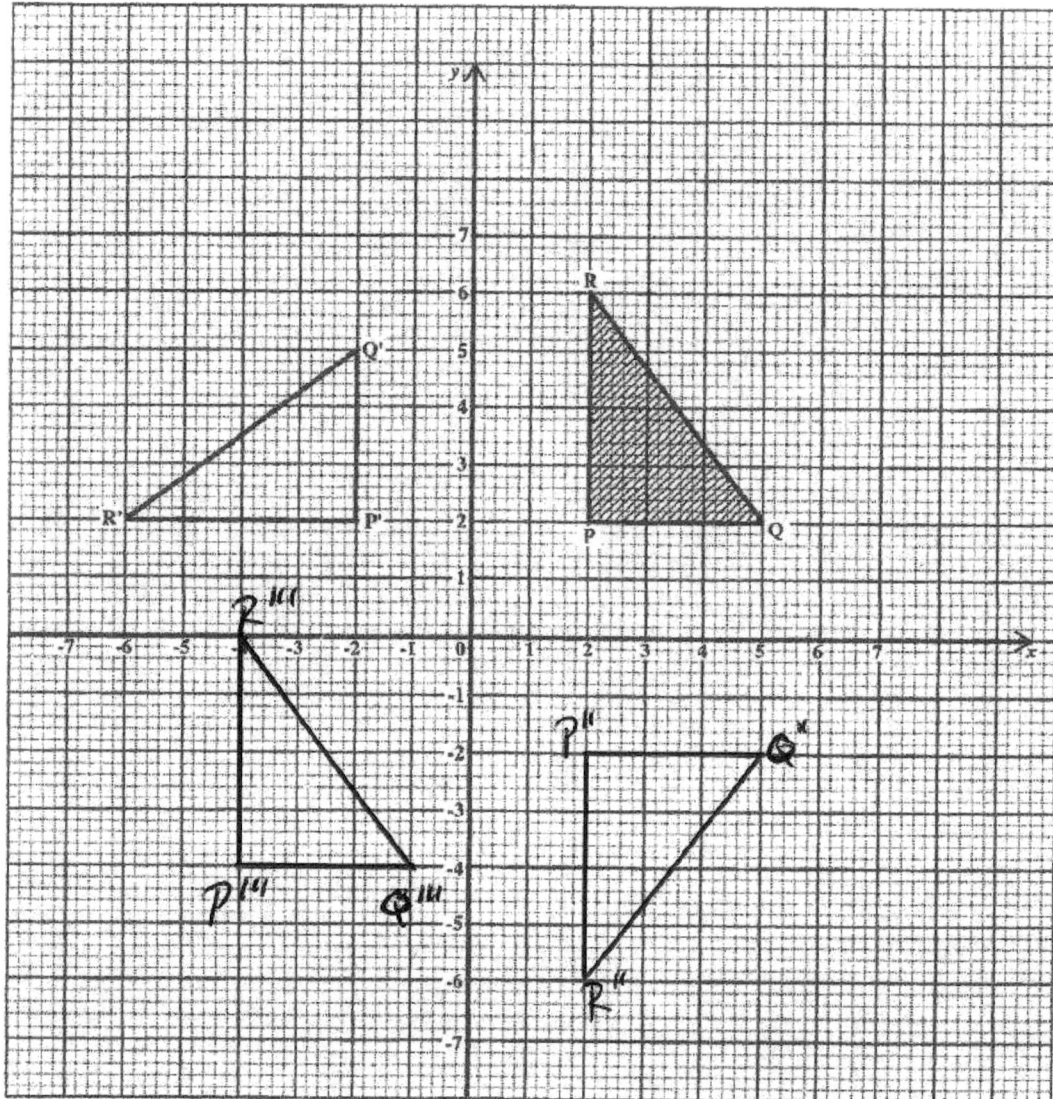

ATTACH THIS ANSWER SHEET TO YOUR ANSWER BOOKLET

000582/F 2003

7 (a) (i) Cost if CD player is bought cash:

$$= 0.95 \times 1200$$
$$= \$1,140$$

(ii) Cost if CD player is bought using Option 1:

$$= \$144 + (\$100 \times 12)$$
$$= \$144 + \$1200$$
$$= \$1,344$$

(iii) Cost if CD player is bought using Option 2:

$$= (0.15 \times 1200) + (\$95 \times 12)$$
$$= \$180 + \$1,140$$
$$= \$1,320$$

(b) Miranda should choose Option 1 because:

(i) The downpayment is less than $150

(ii) The monthly payment is less than $110

8 (a)

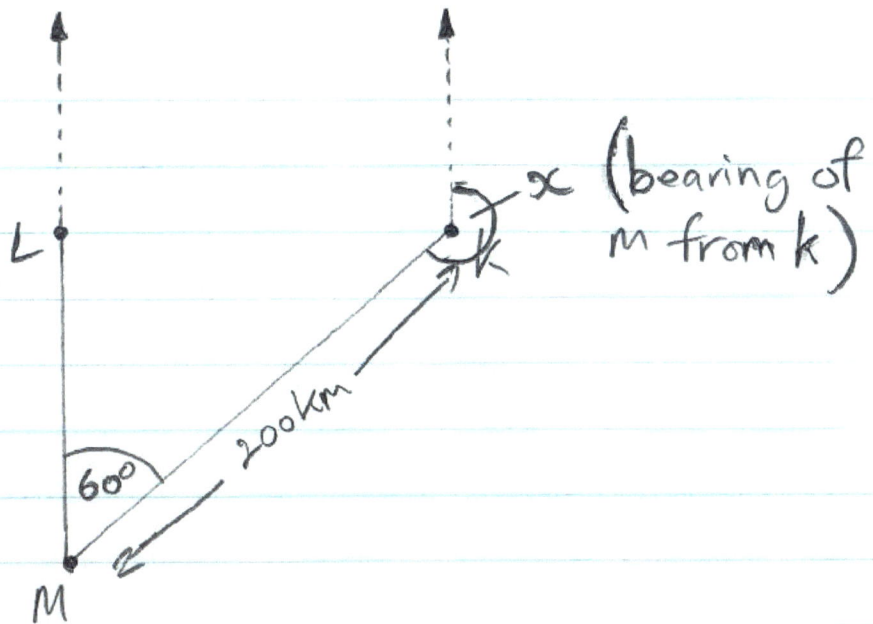

Angle $M\hat{K}K$ is a right angle.

$\therefore \quad \dfrac{LK}{200} = \sin 60$

$\therefore \quad LK = 200 \sin 60$

$\qquad = 173 \text{ km (to the nearest km)}$

9. (a)

Marks	Frequency
10 − 14	2
15 − 19	2
20 − 24	5
25 − 29	8
30 − 34	4
35 − 39	3

(C) P (student scored less than 25 marks)

$$= \frac{\text{No. of students who scored less than 25}}{\text{Total no. of students}}$$

$$= \frac{9}{24}$$

TEST CODE **000582**

FORM TP 23102

MAY/JUNE 2003

CARIBBEAN EXAMINATIONS COUNCIL
SECONDARY EDUCATION CERTIFICATE
EXAMINATION
MATHEMATICS
Paper 02 - Basic Proficiency

Answer sheet for Question 9.   Candidate number ....

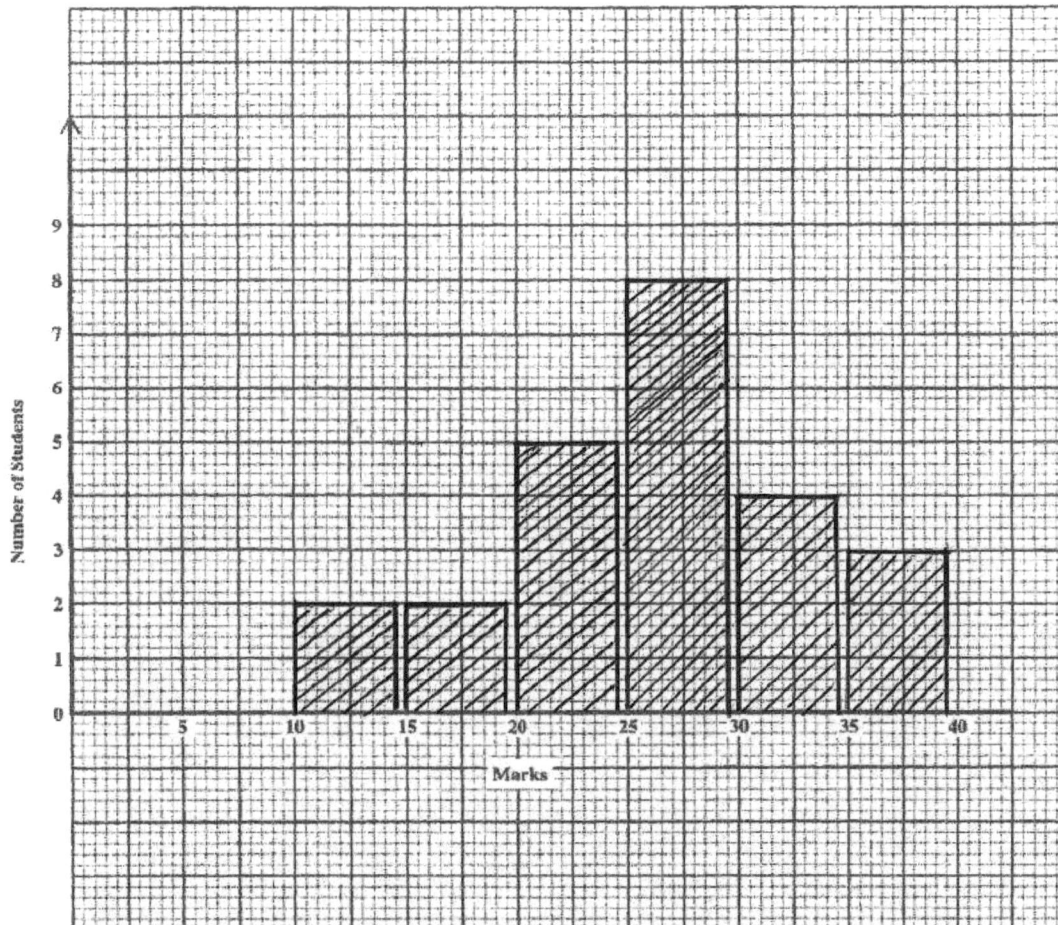

ATTACH THIS ANSWER SHEET TO YOUR ANSWER BOOKLET

000582/F 2003

$$f(x) = x^2 + x - 2$$

19 (a)

x	-3	-2	-1	0	1	2
$f(x)$	4	0	-2	-2	0	4

$$f(-2) = (-2)^2 + (-2) - 2$$
$$= 4 - 2 - 2$$
$$= 4 - 4$$
$$= 0$$

$$f(0) = (0)^2 + (0) - 2$$
$$= 0 + 0 - 2$$
$$= -2$$

$$f(2) = (2)^2 + 2 - 2$$
$$= 4 + 2 - 2$$
$$= 4 + 0$$
$$= 4$$

(d) The two graphs intersect at the points $(-1, -2)$ and $(1, 0)$.

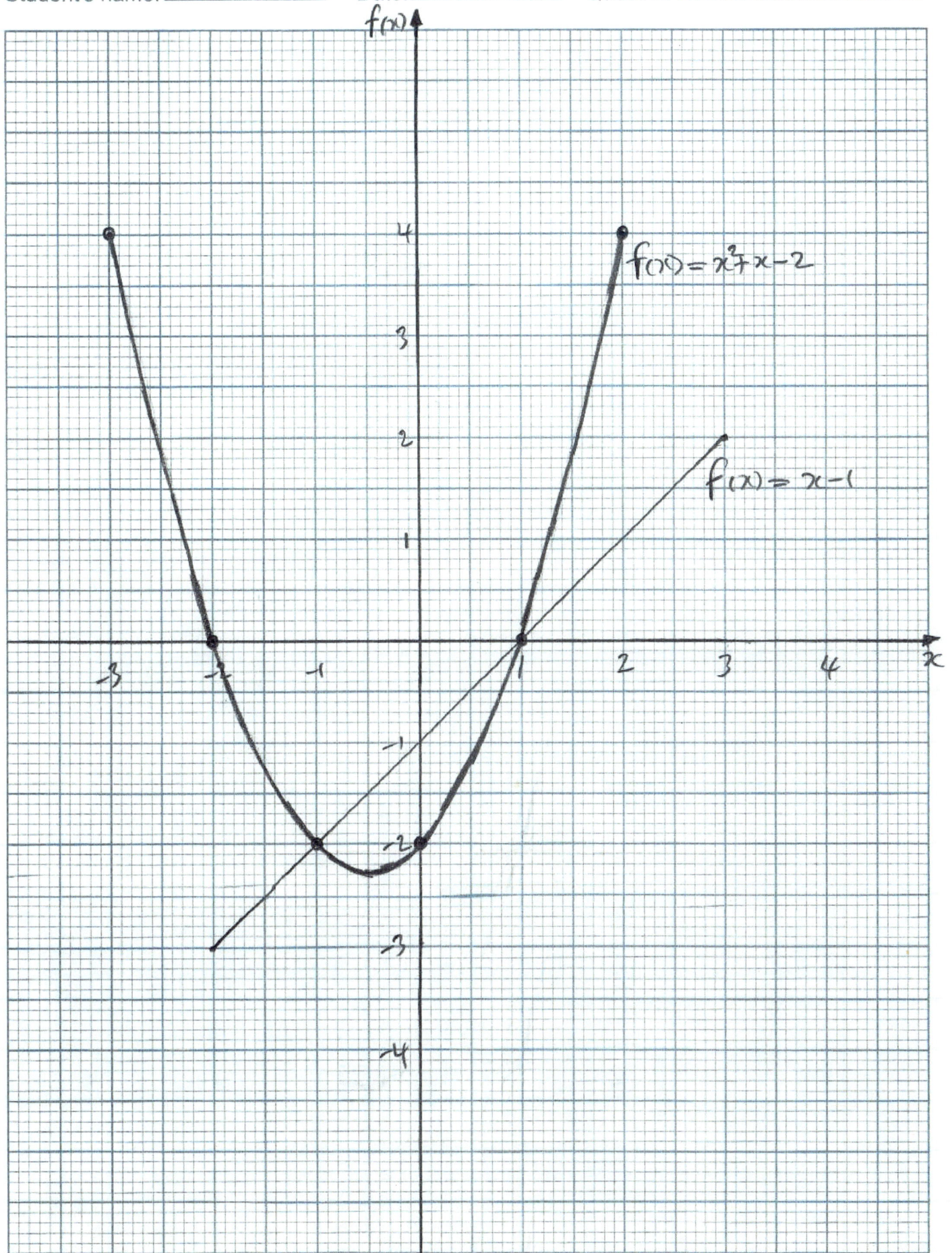

Excelsior

(b)

(c)

$f(x) = x^2 + x - 2$

$f(x) = x - 1$

Excelsior

June 2004

1.(a) $3.48 + \dfrac{3.335}{2.3}$

$\Rightarrow 3.48 + 1.45$

$= 4.93$

$\Rightarrow$ (i) 4.93 (exactly)

(ii) 4.9 (one d.p)

(iii) 5.0 (one s.f)

(b) A ratio of 3:2 means $3+2 = 5$ parts.

$\therefore$ each part $= \dfrac{\$200}{5}$

$= \$40$

$\therefore$ Ratio is $\$40 \times 3 : \$40 \times 2$

$\Rightarrow \$120 : \$80$

(c) 70% of fish was killed. Therefore 30% survived.

150 fish survived $\therefore$ 30% = 150

$$\Rightarrow 1\% = \frac{150}{30}$$

$$= 5$$

$\therefore$ Amt. of fish originally in the pond

$$= 5 \times 100$$

$$= 500$$

(d) $1\frac{3}{4} \div \left(2\frac{1}{2} - 1\frac{1}{3}\right)$

$$\Rightarrow \frac{7}{4} \div \left(\frac{5}{2} - \frac{4}{3}\right)$$

$$\Rightarrow \frac{7}{4} \div \left[\frac{(5)(3) - (4)(2)}{6}\right]$$

$$\Rightarrow \frac{7}{4} \div \left[\frac{15 - 8}{6}\right]$$

$$\Rightarrow \frac{7}{4} \div \frac{7}{6}$$

$$\Rightarrow \frac{7}{4} \times \frac{6}{7}$$

$$\Rightarrow \frac{\cancel{7}}{\cancel{4}_2} \times \frac{\cancel{6}^3}{\cancel{7}}$$

$$= \frac{3}{2}$$

2. (a) $p = 3$; $q = -2$

$$\therefore 4p + 5q = 4(3) + 5(-2)$$
$$= 12 + (-10)$$
$$= 12 - 10$$
$$= 2$$

(b) $\dfrac{x+2}{2} + \dfrac{x-4}{3}$

$$\Rightarrow \frac{3(x+2) + 2(x-4)}{(2)(3)}$$

$$\Rightarrow \frac{3x+6 + 2x-8}{6}$$

$$\Rightarrow \frac{3x+2x+6-8}{6}$$

$$= \frac{5x-2}{6}$$

(c) $6 - 3x \leq 12$

$\Rightarrow \quad -3x \leq 12 - 6$

$\Rightarrow \quad -3x \leq 6$

$\Rightarrow \quad 3x \geq -6$

$\Rightarrow \quad x \geq -\dfrac{6}{3}$

$\Rightarrow \quad x \geq -2$

(d) $3(x - y) - 2(y - x)$

$\Rightarrow \quad 3x - 3y - 2y + 2x$

$\Rightarrow \quad 3x + 2x - 3y - 2y$

$\Rightarrow \quad 5x - 5y$

$\Rightarrow \quad 5(x - y)$

3. (a) (i) Amt. repaid $= 350 \times 18$
$$= \$6,300$$

$\therefore$ Interest paid $= \$6,300 - \$6,000$

$$= \$300$$

(ii) Interest as a percentage of loan

$$= \frac{300}{6000} \times \frac{100}{1}$$

$$= \frac{\overset{5}{\cancel{300}}}{\cancel{6000}} \times \frac{\cancel{100}}{\cancel{1}}$$

$$= 5\%$$

(b)   Cash Price = $696

Hire Purchase Price

$$= (10\% \text{ of } \$696) + (15 \times \$45)$$

$$= (0.1 \times 696) + (15 \times 45)$$

$$= \$69.60 + \$675$$

$$= \$744.60$$

∴ Hire purchase price is more than

cash price by: 744.60 − 696

$$= \$48.60$$

(c)(i) Mrs. Ray worked 3 hrs overtime

∴ She earned $3 \times \$8.40$ in overtime pay

$$= \$25.20$$

∴ Amt. earned at the basic rate

$$\overline{= \#6} \qquad = \$165.20 - \$25.20$$

$$= \$140.00$$

(ii) No. of hours worked at the basic rate

$$= \frac{140}{5.6}$$

$$= 25 \text{ hrs}$$

4. (a)(i) $W = 56°$ (alternate angles)

(ii) $x = 180° - 56°$
$\quad = 124°$

(b) $\angle ABO = 90°$ (angle subtended by a diameter)

(c)(i) Radius of semicircle $= 4cm$

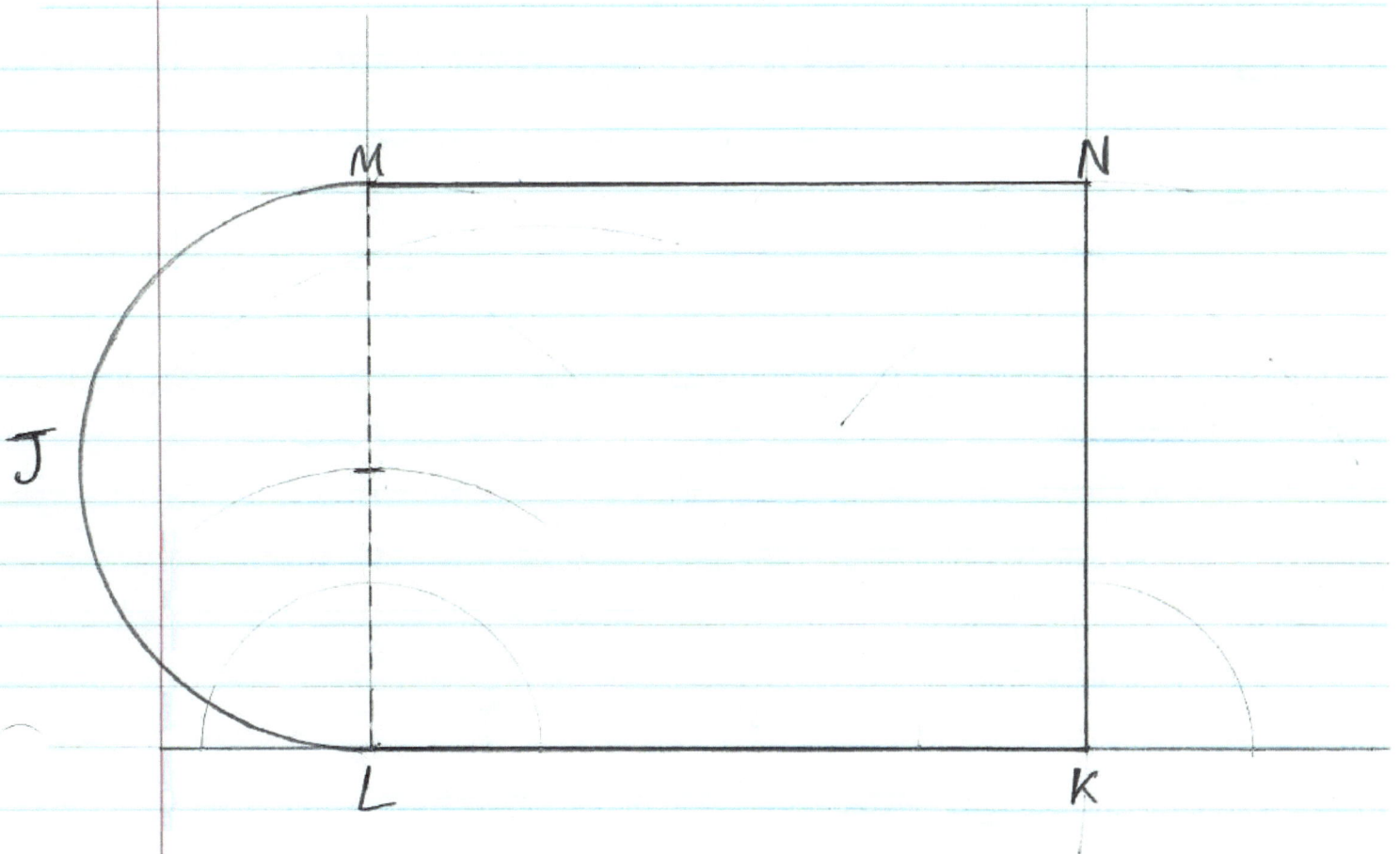

$\underline{\underline{5}}$ (a) (i)   EC$1.00 = US$0.35

$\therefore$ $EC\$\left(\dfrac{1.00}{0.35}\right) = US\$1.00$

$\therefore$ $EC\$\left(\dfrac{1.00}{0.35} \times 6650\right) = US\$6650$

$=$ $EC\$19,000$

$\therefore$ Amt investor paid the bank $= 1\%$ of $\$19,000$

$= 0.01 \times 19,000$

$= \$190$ (EC dollars)

(ii)   EC$1.00 = US$0.38

$EC\$\left(\dfrac{1.00}{0.38}\right) = US\$1$

$\therefore$ $EC\$\left(\dfrac{1.00}{0.38} \times 6650\right) = US\$6650$

$= EC\$17,500$

Amt left for investor in first transaction

$$= \$19,000 - \$190$$
$$= \$18,810$$

Amt left for investor in second transaction

$$= \$17,500 - \$200$$
$$= \$17,300$$

$$\therefore \text{Loss} = \$18,810 - \$17300$$
$$= \$1,510$$

$$\% \text{Loss} = \frac{1510}{18810} \times 100$$

$$= 8.03\%$$

(b) (i) Value after first year $= 1.05 \times 400$
$$= \$420$$

Value after second year $= 1.05 \times 420$
$$= \$441$$

$\therefore$ Value after two years $= \$441$

(ii) Value after third year $= 1.05 \times 441$
$$= \$463.05$$

$\therefore$ Value after three years $= \$463.05$

6 (a) $AD = 5cm + 7cm$

$= 12 cm$

(b) (i) $CD = \frac{1}{4} \times 2\pi r$

$= \frac{1}{4} \times 2 \times \frac{22}{7} \times 7$

$= \frac{1}{4} \times \frac{2}{1} \times \frac{22}{7} \times \frac{7}{1}$

$ \quad 2$

$= \frac{22}{2}$

$= 11 cm$

(ii) Perimeter of ABCDE

$= 7cm + 5cm + 11cm + 7cm + 5cm$

$= 12cm + 18cm + 5cm$

$= 30 cm + 5cm$

$= 35cm$

(iii) Area of ABCDE $= \frac{1}{4}\pi r^2 + (7 \times 5)$

$$= \frac{1}{4} \times \frac{22}{7} \times \frac{7^2}{1} + 35$$

$$= \frac{1}{4} \times \frac{22}{7} \times \frac{49}{1} + 35$$

$$= \frac{1}{\cancel{4}_2} \times \frac{\cancel{22}^{11}}{\cancel{7}} \times \frac{\cancel{49}^7}{1} + 35$$

$$= \frac{77}{2} + 35$$

$$= \frac{77}{2} + \frac{70}{2}$$

$$= \frac{147}{2} \text{ cm}^2$$

$$= 73.5 \text{ cm}^2$$

(c) If the diagram is drawn to a scale of 1:100, the length of rectangle ABCE would be 700 cm, and the width would be 500cm

$\therefore$ Area = 700cm $\times$ 500 cm
= 350,000 cm^2

$\underline{\underline{\mathbf{I}}}$ (a) $x + 2y = 4 \quad - \quad \textcircled{1}$

$\qquad 4x + 3y = 1 \quad - \quad \textcircled{2}$

$\textcircled{1} \times 4: \quad 4x + 8y = 16 \quad -\textcircled{3}$

$\qquad\qquad 4x + 3y = 1 \quad -\textcircled{2}$

$\textcircled{3} - \textcircled{2}: \qquad 5y = 15$

$\qquad\qquad\qquad y = \dfrac{15}{5}$

$\qquad\qquad\qquad = 3.$

Subst. $y = 3$ into eqn $\textcircled{1}$:

$\qquad\qquad x + 2(3) = 4$

$\qquad \Rightarrow \quad x + 6 = 4$

$\qquad\qquad x = 4 - 6$

$\qquad\qquad\quad = -2$

$\qquad \therefore x = -2, \quad y = 3$

(b) (i) $12+x$

(ii) $2+x$

(iii) $12+x = 3(2+x)$

(iv) $12+x = 3(2+x)$
$$12+x = 6+3x$$
$$12-6 = 3x-x$$
$$6 = 2x$$
$$\frac{6}{2} = x$$
$$\therefore x = 3$$

(v) Pearl was paid $3 for the job

8. (a)

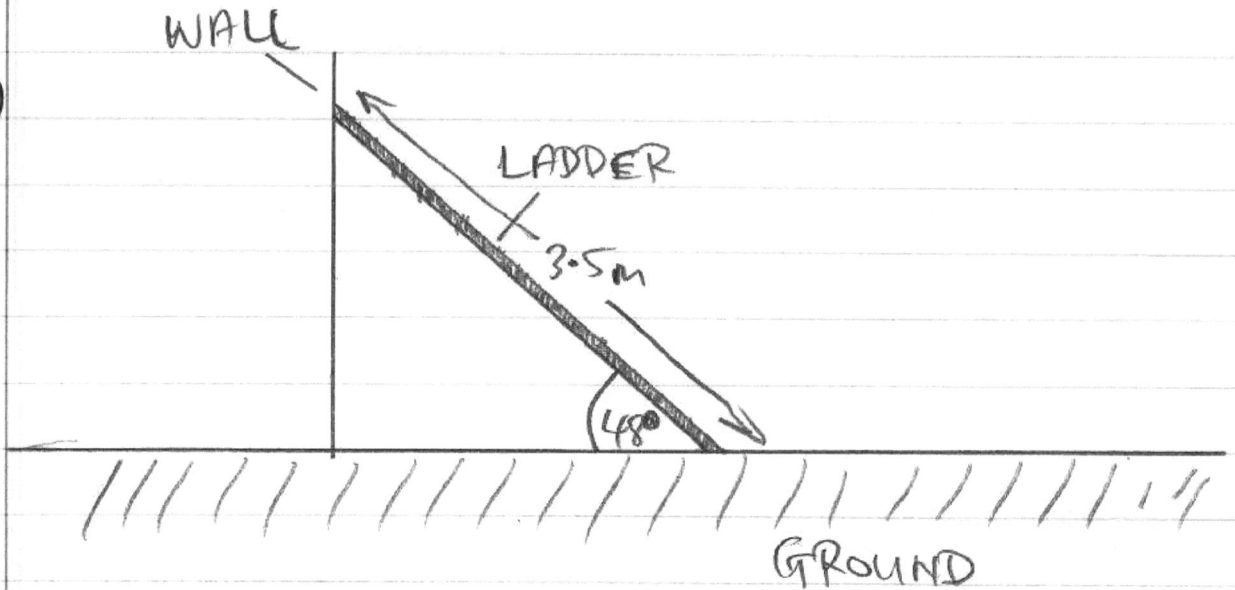

(b) (i) Let height the ladder reaches up the wall
be h.

$$\therefore \quad \frac{h}{3.5} = \sin 48°$$

$$\Rightarrow \quad h = 3.5 \sin 48$$
$$= 3.5 \times 0.7431$$
$$= 2.60 \, m$$

$\therefore$ Height the ladder reaches up the
wall = 2.60m

(ii) Let the distance the foot of the ladder is from the wall be d.

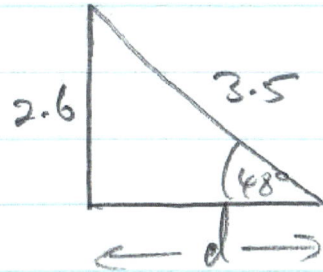

$\therefore$ Using Pythagoras' Theorem:

$$2.6^2 + d^2 = 3.5^2$$

$$d^2 = 3.5^2 - 2.6^2$$
$$= 12.25 - 6.76$$
$$= 5.49$$

$$\therefore d = \sqrt{5.49}$$

$$= 2.34\,m$$

$\therefore$ The foot of the ladder is $2.34\,m$ from the wall.

(c) Let angle the ladder makes with the wall be x.

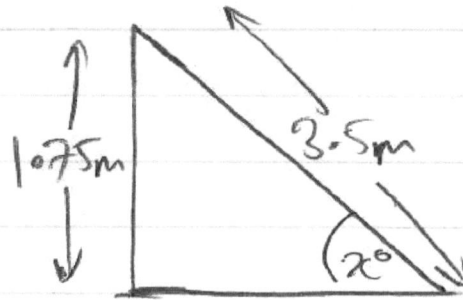

$$\therefore \frac{1.75}{3.5} = \sin x°$$

$$\Rightarrow \quad 0.5 = \sin x°$$

$$x = \sin^{-1} 0.5$$

$$= 30°$$

∴ The ladder now makes an angle of $30°$ with the horizontal.

9 (a)

Marks	Frequency
1	2
2	2
3	4
4	5
5	8
6	3
7	1

(b) (i) The modal mark is 5

(ii) The median mark is 4

(iii) The range is 6   (7 - 1)

(d) P(pupil's mark greater than 5) = $\dfrac{\text{No. of pupils scoring more than 5}}{\text{No. of Pupils}}$

$= \dfrac{4}{25}$

(C)

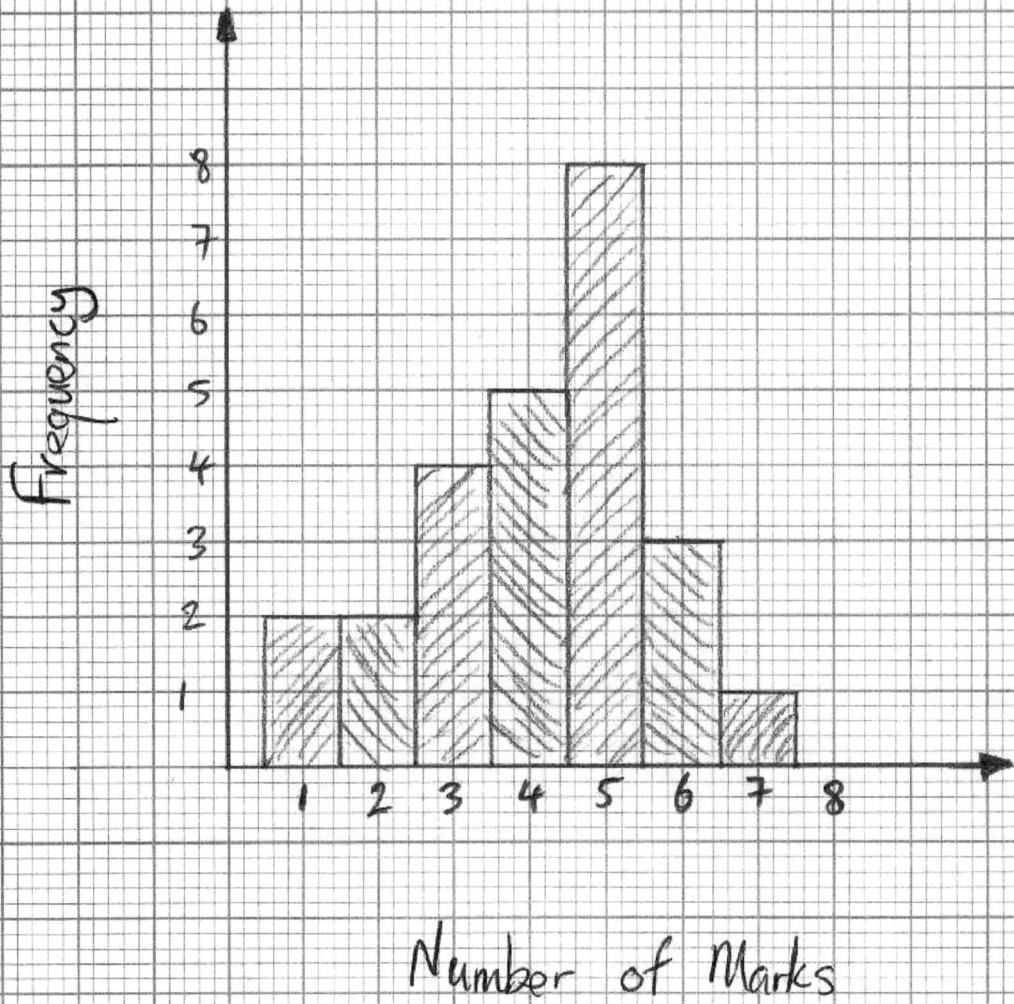

Number of Marks

Excelsior

19 (a) (i) Cost of hiring taxi to travel 250 km

$$= \$70$$

(ii) Cost of hiring taxi to travel 155 km

$$= \$51$$

(b) For a cost of $40, the distance travelled is 100 km

(c) Basic charge = $20

(d) Gradient of line $= \dfrac{30-20}{50-0} = \dfrac{10}{50}$

$$= 0.2$$

(e) Equation of line: $y = 0.2x + 20$

(f) Cost of hiring taxi to travel 330 km
$$= 0.2(330) + 20$$
$$= 66 + 20$$
$$= \$86$$

June 2005

1. (a) (i) $\dfrac{1.092}{24} = 0.0455$ (exactly)

(ii) $= 4.55 \times 10^{-2}$ (standard form)

(b) Let x be the number of students in the school.

58% of students are girls. There are 406 girls in the school.

∴ 58% of x ~~are~~ are girls

∴ $0.58x = 406$

$x = \dfrac{406}{0.58}$

$= 700$

∴ There are 700 students in the school.

(c) (i) $72 : x = 4 : 3$

$\Rightarrow$ $\dfrac{72}{x} = \dfrac{4}{3}$

Cross-multiply: $72(3) = 4x$

$$216 = 4x$$
$$\dfrac{216}{4} = x$$

$$54 = x$$

$\therefore$ Team B scored 54 points in the
second round

(ii) Team A total score $= 96 + 72$
$$= 168$$

For Team B's score to be equal to Team A's
se total score, they would need to
score $168 - 58 = 110$ points.
$\therefore$ Team B would need to score 111 points to
be in the second round to have a

greater total score than Team A's
total score.

2. (a) | Interest at end of first year = 10% of \$800
$$= 0.1 \times 800$$
$$= \$80$$

Interest at end of second year = 10% of
$$(800 + 80)$$
$$= 10\% \text{ of } \$880$$
$$= 0.1 \times 880$$
$$= \$88$$

∴ Interest earned over two years
$$= \$80 + \$88$$
$$= \$168$$

(b) (i) Basic weekly wage $= 40 \times 16$

$\qquad = \$640$

(ii) Wage for 47 hour work week

$= $ Basic weekly wage $+$ Wage for 7 hours overtime

$$\left[\begin{array}{l} \text{Overtime} = 47 - 40 \\ \qquad\qquad = 7 \text{ hrs} \end{array} \right]$$

Overtime hourly rate $= \$16 + \$4$

$\qquad\qquad = \$20$

$\therefore$ Wage for 7 hrs overtime $= 20 \times 7$

$\qquad\qquad\qquad = \$140$

$\therefore$ Wage for 47 hour work week

$\qquad\qquad = 640 + 140$

$\qquad\qquad = \$780$

(iii) Overtime wage = $860 − $640
       = $220

∴ No. of hours overtime = $\frac{220}{20}$

                        = 11

∴ Number of hours worked = 40 + 11
                         = 51 hrs.

3 (a) (i) Total annual tax-free allowances

$$= (\$900 \times 2) + (\$400 \times 2) + \$2500$$

$$= \$1800 + \$800 + \$2500$$

$$= \$2600 + \$2500$$

$$= \$5100$$

(ii) Annual taxable income = Annual income − Total tax-free allowances

$$= \$32,000 - \$5,100$$

$$= \$26,900$$

(iii) Annual taxes $= \$1,200 + 30\% \text{ of } (26,900 - 20,000)$

$$= \$1,200 + (0.3 \times 6,900)$$

$$= \$1,200 + \$2,070$$

$$= \$3,270$$

(b) (i) Cash price = $8,400 − (12% of 8,400)

$= \$8,400 - (0 \cdot 12 \times 8,400)$

$= \$8,400 - \$1,008$

$= \$7,392$

(ii) Hire purchase price = $\$2,940 + (24 \times 230)$

$= \$2,940 + \$5,520$

$= \$8,460$

(iii) Amt. saved by buying cash

$= 8460 - 7392$

$= \$1,068$

4. (a) $4x + 3y = 26$ — ①
         $2x - y = 8$ — ②

         $4x + 3y = 26$ — ①
② × 2: $4x - 2y = 16$ — ③

① − ③: $3y - (2y) = 26 - 16$

         $3y + 2y = 10$
             $5y = 10$
              $y = \dfrac{10}{5}$
              $y = 2$

Subst. $y = 2$ into eqn ②:

         $2x - 2 = 8$
           $2x = 8 + 2$
              $= 10$
            $x = \dfrac{10}{2}$
            $x = 5$

         $\therefore x = 5, \ y = 2$

(b) (i) Angle $C = (p+3)^\circ$

(ii) $p + p+3 + p+3 = 180^\circ$

$\Rightarrow \quad p+p+p+3+3 = 180$

$\Rightarrow \qquad\qquad 3p + 6 = 180$

$\Rightarrow \qquad\qquad\quad 3p = 180 - 6$

$\qquad\qquad\qquad\qquad = 174$

$\Rightarrow \qquad\qquad\quad p = \dfrac{174}{3}$

$\qquad\qquad\qquad\qquad = 58^\circ$

$\therefore$ Angle $A = 58^\circ$
Angle $B = 58+3 = 61^\circ$
Angle $C = 58+3 = 61^\circ$

5 (a) $x = 2$, $y = -3$

$\therefore xy^2 = 2(-3)^2$

$\qquad = 2(9)$

$\qquad = 18$

(b) $\dfrac{3x}{12} - \dfrac{x+2}{8}$

$\Rightarrow \dfrac{3x(8) - 12(x+2)}{(12)(8)}$

$\Rightarrow \dfrac{24x - 12x - 24}{96}$

$\Rightarrow \dfrac{12x - 24}{96}$

$\Rightarrow \dfrac{12(x-2)}{96} = \dfrac{x-2}{8}$

(c)(i) $3x - 5 - 5x < 7$

$\Rightarrow$ $3x - 5x - 5 < 7$

$\Rightarrow$ $-2x - 5 < 7$

$\Rightarrow$ $-2x < 7 + 5$

$\Rightarrow$ $-2x < 12$

$\Rightarrow$ $2x > -12$

$\Rightarrow$ $x > -\dfrac{12}{2}$

$\Rightarrow$ $x > -6$

(ii)

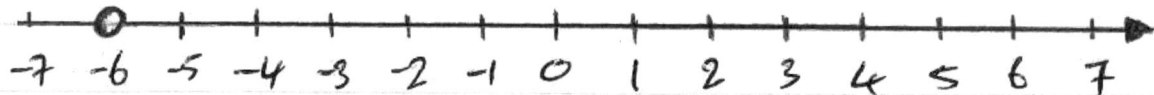

6. (a) (i) Gradient of $l = \dfrac{2-(-1)}{3-0}$

$$= \dfrac{2+1}{3}$$

$$= \dfrac{3}{3}$$

$$= 1$$

(ii) Equation of l:

$$y = x - 2 \qquad \left[\begin{array}{l} \text{gradient} = 1 \\ y\text{-intercept} = -2 \end{array}\right]$$

(b) (i)

x	0	1	2	3	4
$f(x)$	3	0	-1	0	3

$f(x) = x^2 - 4x + 3$

$f(1) = (1)^2 - 4(1) + 3$
$\quad = 1 - 4 + 3$
$\quad = -3 + 3$
$\quad = 0$

$f(4) = (4)^2 - 4(4) + 3$
$\quad = 16 - 16 + 3$
$\quad = 0 + 3$
$\quad = 0$

(iii) From the graph, the line l and the graph of $f(x)$ intersect at the point $(1, 0)$

CARIBBEAN EXAMINATIONS COUNCIL

SECONDARY EDUCATION CERTIFICATE
EXAMINATION

MATHEMATICS

Paper 02 – Basic Proficiency

Answer Sheet for Question 6 **(b) (ii)**   **Candidate Number** ....

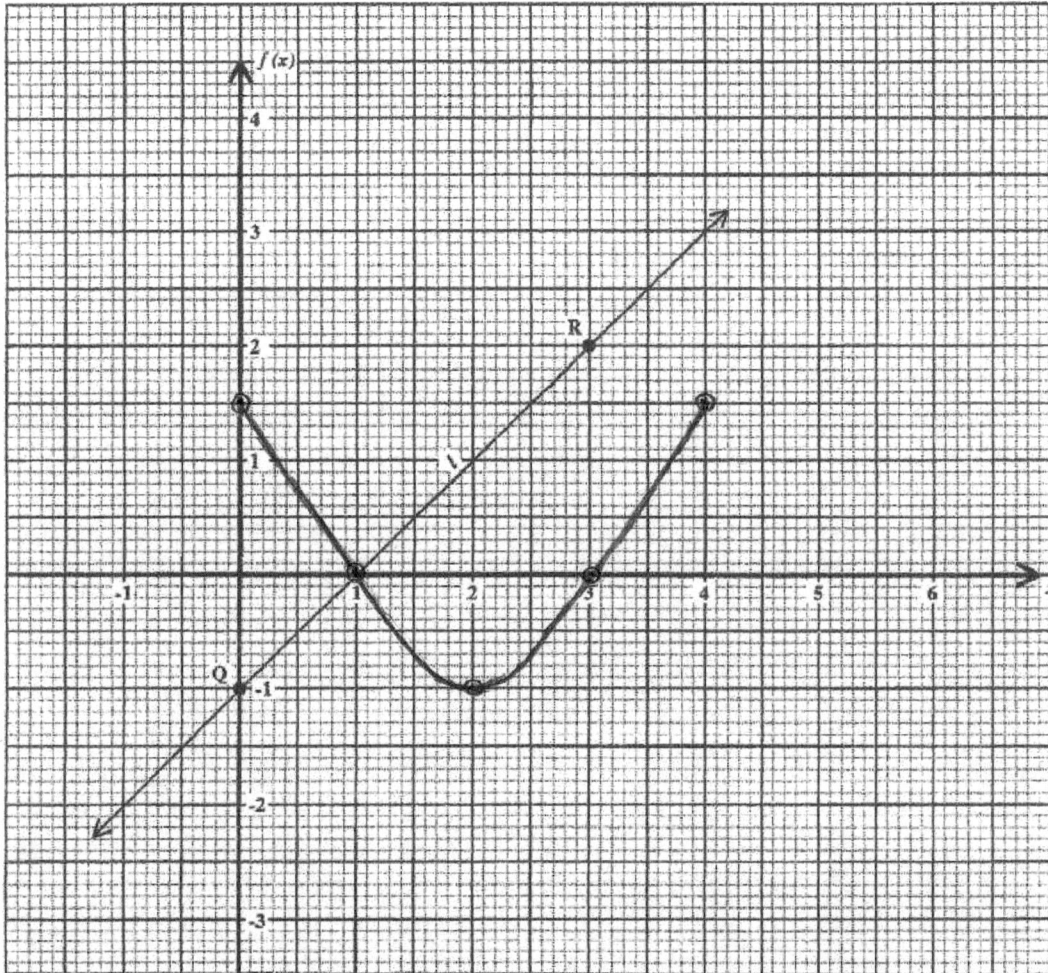

ATTACH THIS ANSWER SHEET TO YOUR ANSWER BOOKLET

01134020/F 2005

7 (a) (i)   07:30 a.m to 08:00 a.m = 30 minutes

08:00 a.m to 08:10 a.m = 10 minutes

∴ Time taken by the bus to travel from Town X to Town Y = 30 + 10
= 40 minutes

(ii) Average Speed $= \dfrac{Distance}{Time}$

$= \dfrac{32km}{40\ minutes}$

40 minutes $= \dfrac{40}{60}$ hrs

$= \dfrac{2}{3}$ hrs

∴ Ave. speed in kmh⁻¹ $= \dfrac{32}{2/3} = \dfrac{32}{1} \times \dfrac{3}{2}$

$\dfrac{\overset{16}{\cancel{32}}}{1} \times \dfrac{3}{\cancel{2}_1} = \dfrac{48}{1}$

$= 48\ kmh^{-1}$

(b)(i) Area of base of container = 75 × 40

= 3,000 cm²

(ii) Volume of Water = Area of Base × Height of Water

= 3000 × 15

= 45,000 cm³

(iii) Volume of container when full = 84 litres

= 84,000 cm³

∴ 84,000 = Area of Base × Height of Container

= 3,000 × h

∴ 84,000 = 3,000h

⟹ $\frac{84,000}{3,000}$ = h

⟹ 28 = h

∴ Height of water when container is full

= 28 cm

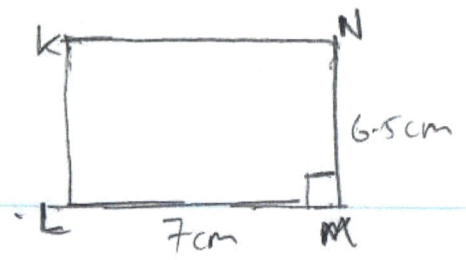

K — N
6.5cm
L — M
7cm

K

N

L

M

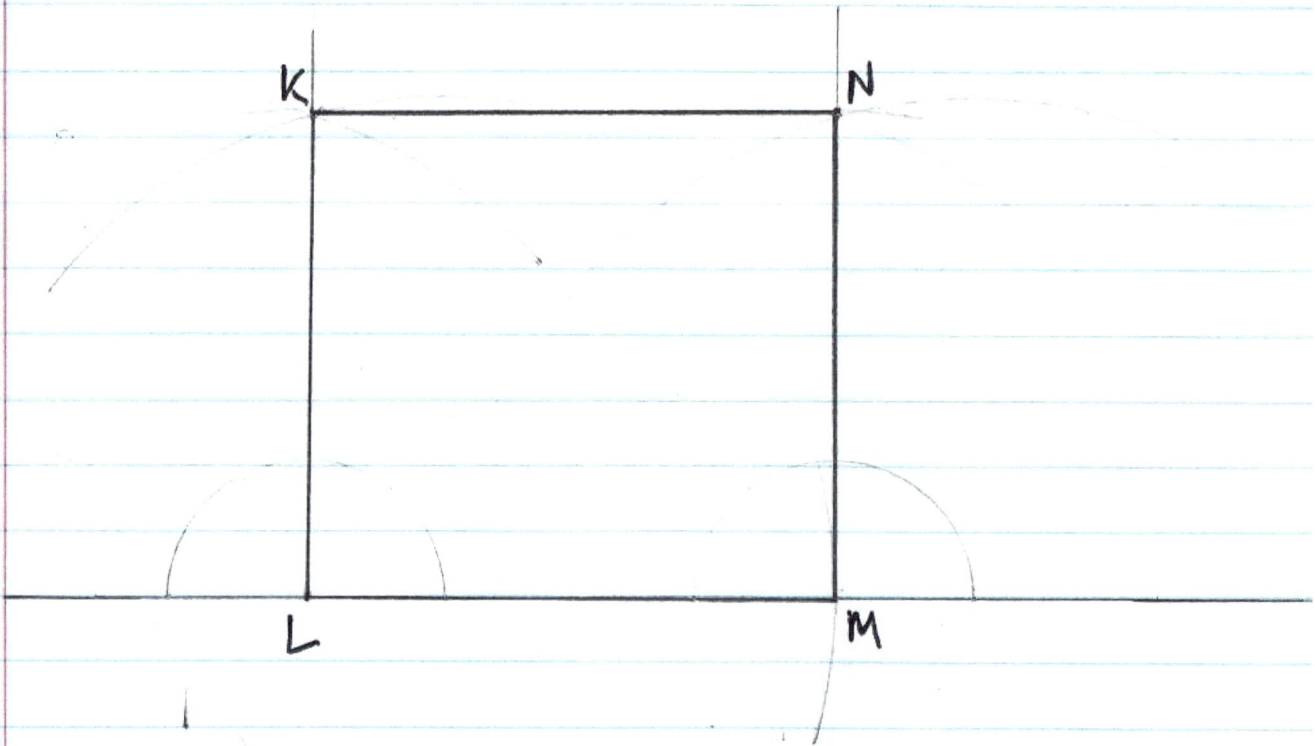

(ii) Diagonal LN = 10.5 cm

(b) (i) $12^2 + 16^2 = BD^2$ (Pythagoras' Theorem)

$\Rightarrow$ $144 + 256 = BD^2$

$400 = BD^2$

$\therefore$ $BD = \sqrt{400}$

$= 20\,cm$

$BD = 20\,cm$

(ii) $\sin M = \dfrac{GF}{10}$

$= \dfrac{8}{10}$

$\Rightarrow$ $\sin M = 0.8$

$M = \sin^{-1} 0.8$

$= 53.1°$

$\therefore$ Angle of elevation of G from M $= 53.1°$

9 (a) $\begin{pmatrix} -2 \\ 7 \end{pmatrix} + \begin{pmatrix} x \\ y \end{pmatrix} = \begin{pmatrix} 8 \\ 23 \end{pmatrix}$

$\therefore \quad -2 + x = 8$

$\qquad x = 8 + 2$

$\qquad = 10$

$\Rightarrow \quad 7 + y = 23$

$\qquad y = 23 - 7$

$\qquad = 16$

$\therefore$ Column vector $= \begin{pmatrix} 10 \\ 16 \end{pmatrix}$

$\qquad x = 10, \quad y = 16$

(b) (i) $BC^2 + CD^2 = BD^2$

$\Rightarrow$ $\quad 12^2 + 16^2 = BD^2$

$\Rightarrow$ $\quad 144 + 256 = BD^2$

$\Rightarrow$ $\qquad 400 = BD^2$

$\Rightarrow$ $\quad \sqrt{400} = BD$

$\Rightarrow$ $\qquad 20 = BD$

$\Rightarrow$ $\qquad BD = 20 \text{ cm}$

(ii) Scale factor of enlargement $= \dfrac{AE}{BD}$

$$= \dfrac{60}{20}$$

$$= 3$$

10. (a) 12 children wear a size 7 shoe

(b) No. of children who wear a shoe size
   smaller than size 7 = 10 (size 6)
                        +14 (size 5) + 5 (size 4)

$$= 10 + 14 + 5$$
$$= 24 + 5$$
$$= 29$$

(c) The modal shoe size is size 5.

(d) The median shoe size is size 6.

(e)(i) $P(\text{size } 5) = \dfrac{14}{50} = 0.28$

(ii) $P(\text{size larger than } 6) = \dfrac{12+9}{50}$

$$= \dfrac{21}{50} = 0.42$$

(f) ~~The median would be of interest to the
shop owner only if its real value wa~~

(f) The owner of the shoe store would be interested in which shoe size sells the most, and not in which shoe size happened to be at the centre of the distribution of shoe sizes.

The mode would therefore be of greater interest to the shoe store owner.

June 2006

$\underline{1}$ (a) $\dfrac{0.6}{7.5} = \dfrac{6}{75} = \dfrac{2}{25}$

(b) $\dfrac{1\frac{5}{8} + \frac{1}{4}}{\frac{3}{4}} = \dfrac{\frac{13}{8} + \frac{1}{4}}{\frac{3}{4}}$

$$= \dfrac{\frac{13}{8} + \frac{2}{8}}{\frac{3}{4}}$$

$$= \dfrac{\frac{15}{8}}{\frac{3}{4}}$$

$$= \dfrac{15}{8} \times \dfrac{4}{3} \qquad\qquad \dfrac{\cancel{15}^{5}}{8} \times \dfrac{\cancel{4}^{1}}{\cancel{3}_{2}} \cdot = \dfrac{5}{2}$$

$$= \dfrac{5}{2}$$

(c) (i) US $85 $= (85 \times 2)$ Barbados dollars
              $=$ BDS $170

(ii) Barbados dollars remaining $= 170 - 94$
                                $=$ BDS $76

BDS $76 $= \dfrac{76}{2}$ US Dollars

       $=$ US $38

2. (a) $\dfrac{x^5 \times x^4}{x^3} = \dfrac{x^{5+4}}{x^3} = \dfrac{x^9}{x^3} = x^{9-3}$

$$= x^6$$

(b) $m * n = (m-n)^2$

(i) $5 * 2 = (5-2)^2 = 3^2 = 9$

(ii) $2 * 5 = (2-5)^2 = (-3)^2 = 9$

$$\therefore 2 * 5 = 5 * 2$$

(c) $\begin{array}{ll} 3x + 2y = 17 & \text{①} \\ x - 2y = 3 & \text{②} \end{array}$

① + ② : $4x = 20$

$x = \dfrac{20}{4}$

$= 5$

Substitute $x = 5$ into eqn ② :

$5 - 2y = 3$

$5 - 3 = 2y$

$2 = 2y$

$y = \frac{2}{2} = 1$

$\therefore x = 5, y = 1$

3 (a) Value of car at the end of the first year:

$$= \$40,000 - (10\% \text{ of } \$40,000)$$

$$= \$40,000 - (0.1 \times 40,000)$$

$$= \$40,000 - \$4,000$$

$$= \$36,000$$

Value of car at the end of second year:

$$= \$36,000 - (10\% \text{ of } \$36,000)$$

$$= \$36,000 - (0.1 \times 36,000)$$

$$= \$36,000 - \$3,600$$

$$= \$32,400$$

$\therefore$ Value of car after two years

$$= \$32,400$$

(b)(i) Time taken to save enough money to buy watch $= \dfrac{78}{6}$

$$= 13 \text{ weeks}$$

(ii) Total hire purchase price $= 4 + (15 \times 5.50)$

$$= 4 + 82.5$$
$$= \$86.50$$

Daveed should buy the watch for cash. It is cheaper, and it takes less time

4 (a) (i) Length of side $= \dfrac{20.8}{4}$

$= 5.2$ cm

(ii) Area of square $= 5.2 \times 5.2$

$= 27.04$ cm^2

(b) (i) Volume of tank = Area of base $\times$ Height

$= 750$ cm^2 $\times$ 120 cm

$= 90,000$ cm^3

(ii) 54 litres $= 54,000$ cm^3

(iii) Percentage of tank water occupies

$= \dfrac{54000}{90000} \times \dfrac{100}{1}$

$= 60\%$

$\underline{\underline{5}}$ (a) $\quad p = -3 \qquad q = 5$

$$\frac{q-p}{q+p} = \frac{5-(-3)}{5+3}$$

$$= \frac{5+3}{5+3}$$

$$= \frac{8}{8}$$

$$= 1$$

(b) $\dfrac{x+1}{2} - \dfrac{x}{3} = \dfrac{3(x+1) - 2x}{(2)(3)}$

$$= \frac{3x+3 - 2x}{6}$$

$$= \frac{3x - 2x + 3}{6}$$

$$= \frac{x+3}{6}$$

(c) (i) a) Pam's age: $x+12$ years old

b) Sum of Andy's age and Pam's age:

$$x + x + 12$$

$$= 2x + 12$$

(ii) ??

Info missing for these two parts

(iii) ??

6, (a) (i) Total cost of typing document with 10 pages:

$$= 40 \text{ cents} \times 10$$

$$= 400 \text{ cents}$$

$$= \$4.00$$

(ii) Total cost of typing document with 15 pages:

$$= 40 \text{ cents} \times 15$$

$$= 600 \text{ cents}$$

$$= \$6.00$$

(iii) Total cost of typing document with 23 pages:

$$= (40 \text{ cents} \times 15) + (30 \text{ cents} \times 8) \qquad [23 - 15 = 8]$$

$$= 600 + 240$$

$$= 840 \text{ cents}$$

$$= \$8.40$$

(b) $18.00 = 1800$ cents

$\therefore 1800 = (40 \text{ cents} \times 15) + (30 \text{ cents} \times n)$

Let n be the number of pages in excess of the first 15 pages.

$\therefore 1800 = (40 \times 15) + 30n$

$1800 = 600 + 30n$

$1800 - 600 = 30n$

$1200 = 30n$

$\dfrac{1200}{30} = n$

$n = 40$

$\therefore$ The document contains $(40 + 15)$ pages

$= 55$ pages

7. (a) (i)

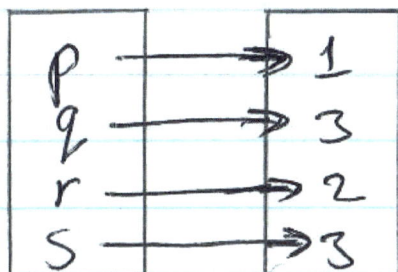

(ii) The mapping is many-to-one because both q and s map to 3.

(b) (i) The temperature at the start of the experiment was 25°C.

(ii) Highest temperature – 35°C
Lowest temperature – 20°C

∴ Difference = 35 – 20
            = 15°C

(iii) Rate of decrease in temperature

= $\dfrac{\text{Decrease in temperature}}{\text{Time taken}}$ = $\dfrac{15°C}{2\,hr}$

= 7.5°C/hr

(iv) The temperature was rising most quickly between the one-hour period 4pm to 5pm.

8. (a)

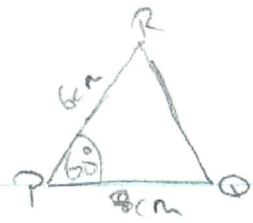

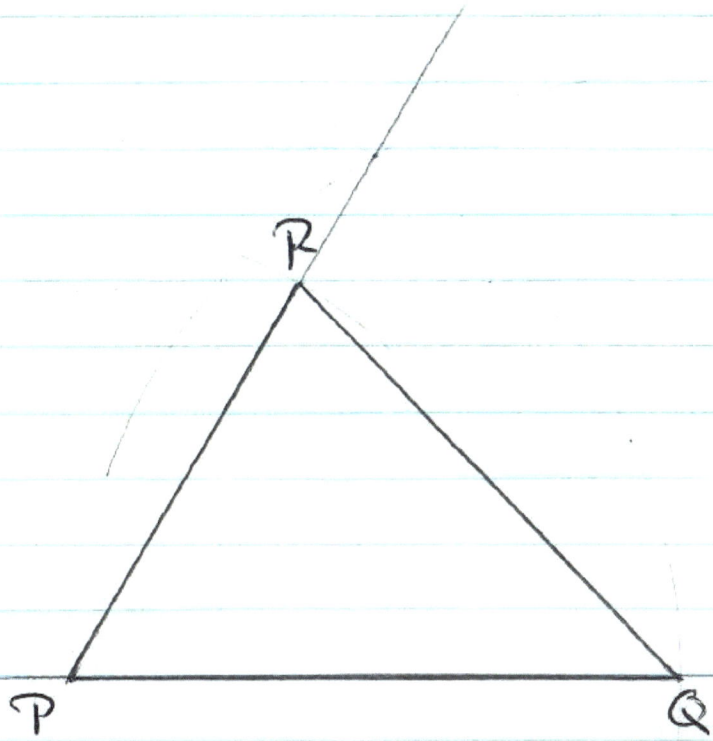

QR = 7.2 cm

(b)(i) The rotation is 90° clockwise. The centre
of rotation is the origin

(ii) a) The length of A'B' will remain unchanged
under a ~~translatic~~ translation.
∴ A"B" = 6 units

b) The angle ∠'A'B' will remain unchanged under
a translation. ∴ ∠C"A"B" = 90°

9 (a) 15 more chocolates were sold on Friday than on Tuesday $(60-45=15)$

(b) Total number of chocolates sold for the week $= 22 + 45 + 58 + 35 + 60$
$$= 220$$

(c) Mean number of chocolates sold daily

$$= \frac{220}{5}$$

$$= 44$$

CARIBBEAN EXAMINATIONS COUNCIL

SECONDARY EDUCATION CERTIFICATE
EXAMINATION

MATHEMATICS

Paper 02 – Basic Proficiency

Answer Sheet for Question 9 (d)   Candidate Number ....

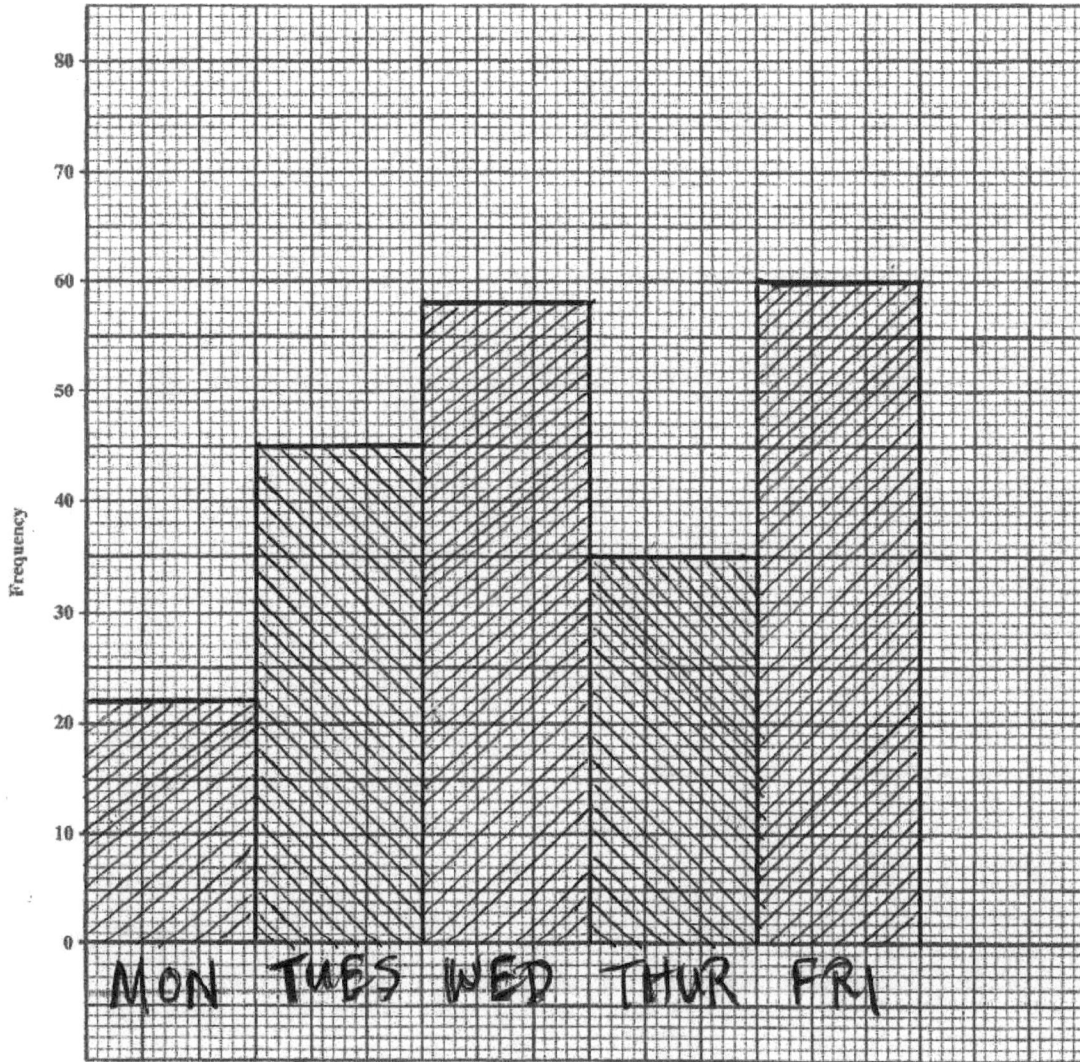

ATTACH THIS ANSWER SHEET TO YOUR ANSWER BOOKLET

01134020/F 2006

(e) Probability that on a day chosen at random, less than 50 chocolates were sold =

$$\frac{\text{No. of days on which less than 50 chocolates sold}}{\text{No. of days}}$$

$$= \frac{3}{5}$$

10 (a) $x + 70 = 180$

$x = 180 - 70$

$= 110°$

y and x are alternate angles

$\therefore \quad y = x$

$y = 110°$

(b) (i) $EH^2 = EF^2 + AF^2$

$17^2 = EF^2 + 8^2$

$289 = EF^2 + 64$

$289 - 64 = EF^2$

$225 = EF^2$

$\sqrt{225} = EF$

$15 = EF$

$\therefore EF = 15m$

(ii) $\tan HGF = \dfrac{8}{5}$

$$= 1.6$$

$\therefore HGF = \tan^{-1} 1.6$

$$= 58°$$

(iii) Bearing of G from F = angle HFG

$$= 90 - 58$$

$$= 32°$$

June 2007

1. (a) $624 \times 1.05^2 = 624 \times 1.1025$

$$= 687.96$$

$$= \text{(i) } 687.96 \text{ (exactly)}$$
$$\text{(ii) } 690 \quad (2 \text{ s.f})$$

(b) Ratio = 2:5
Andy = 2 parts, Bob = 5 parts

Let Andy's share $= x$
∴ Bob's share $= x + 45$

$$\Rightarrow \quad \frac{x}{x+45} = \frac{2}{5}$$

Cross-multiply: $5x = 2(x+45)$
$$5x = 2x + 90$$
$$5x - 2x = 90$$
$$3x = 90$$
$$x = \frac{90}{3}$$

$$x = 30$$

Total amt. of money shared $= x + x + 45$
$$= 30 + (30 + 45)$$

$$= 30 + 75$$
$$= 105$$

∴ Total amt. of money shared = $105

(c) $\frac{1}{4}$ of students play netball

$\frac{3}{8}$ of students play cricket

(i) Fraction of students who play netball and cricket:
$$= \frac{1}{4} + \frac{3}{8}$$

$$= \frac{8}{32} + \frac{12}{32}$$

$$= \frac{20}{32}$$

$$= \frac{5}{8}$$

∴ $\frac{5}{8}$ of students play netball and cricket

(ii) Fraction of students who play basketball:

$$= 1 - \frac{5}{8}$$

$$= \frac{8}{8} - \frac{5}{8}$$

$$= \frac{3}{8}$$

$\therefore$ Percentage of students who play basketball:

$$= \frac{3}{8} \times \frac{100}{1} = \frac{300}{8}$$

$$= 37.5\%$$

2/(a)(i) $2p^2 \times 3p^3 = 2 \times 3 \times p^2 \times p^3$

$\qquad\qquad\qquad = 6 \times p^2 \times p^3$

$\qquad\qquad\qquad = 6p^5$

(ii) $\dfrac{p^3}{p} = \dfrac{p^3 \, p^2}{p}$

$\qquad\qquad = p^2$

(iii) $3(2x+1) - 4x$

$\Rightarrow \quad 6x + 3 - 4x$

$\Rightarrow \quad 6x - 4x + 3$

$\Rightarrow \quad 2x + 3$

(b) $\quad a * b = ab^2$

$\therefore \; 4 * (-5) = 4(-5)^2$

$\qquad\qquad\quad = 4(25)$

$\qquad\qquad\quad = 100$

(c) (i) $8 + 2x > 2 + 5x$

$\Rightarrow \quad 8 - 2 > 5x - 2x$

$\Rightarrow \quad 6 > 3x$

$\Rightarrow \quad \dfrac{6}{3} > x$

$\Rightarrow \quad 2 > x$

$\Rightarrow \quad x < 2$

(ii) If x is a whole number, then the largest possible value of x' is **1**.

3. (a) Cash price = \$340

Hire purchase price = \$15 + 18 × \$20
$$= \$15 + \$360$$
$$= \$375$$

∴ Difference between cash price and
hire purchase price = \$375 − \$340
$$= \$35$$

(b) Total insurance = Land insurance + House insurance

$$= \$100 + 0.5\% \text{ of } \$250,000$$

$$= \$100 + \left(\frac{0.5}{100} \times \frac{250,000}{1}\right)$$

$$= \$100 + \left(\frac{0.5}{\cancel{100}} \times \frac{250\cancel{000}}{1}\right)$$

$$= \$100 + (0.5 \times 2,500)$$
$$= \$100 + \$1,250$$
$$= \$1,350$$

(c) Amt. after first year $= 6000 + (5\% \text{ of } 6000)$

$= 6000 + (0.05 \times 6000)$

$= 6000 + 300$

$= 6300$

∴ Amt. after first year $= \$6,300$

Amt. after second year $= 6300 + (5\% \text{ of } 6300)$

$= 6300 + (0.05 \times 6300)$

$= 6300 + 315$

$= 6615$

∴ Amt. after second year $= \$6,615$

∴ Am ∴ Total amount in account after
two years $= \$6,615$

4 (a) (i) Actual distance from park to church

$$= 5.6 \times 2$$
$$= 11.2 \text{ km}$$

(ii) Actual distance from park to school = 15 km

$\therefore$ Distance on map $= \dfrac{15}{2}$

$$= 7.5 \text{ cm}$$

(b)(i) QT = 14cm − 8cm

$$= 6cm$$

(ii) a) Area of QRT $= \dfrac{6 \times 8}{2} = \dfrac{48}{2}$

$$= 24 \text{ cm}^2$$

b) Area of PTRS = Area of QRT + Area of PQRS

$$= 24 + (8 \times 8) = 24 + 64$$
$$= 88 \text{ cm}^2$$

5 (a) (i) Hourly rate $= \dfrac{\text{Wage}}{\text{Hours}} = \dfrac{360}{45}$

$$= \$8 \text{ per hour}$$

(ii) Overtime rate $= 2 \times 8 = \$16 \text{ per hour}$

(iii) Total wage = Basic wage for 45-hour Week + Overtime wage

$$= \$360 + \text{Overtime wage}$$

Overtime hours $= 50 - 45$
$$= 5 \text{ hrs}$$

$\therefore$ Overtime wage $= 5 \times 16$
$$= \$80$$

$\therefore$ Total wage $= \$360 + \$80$
$$= \$440$$

(b) (i) a) Total cost of books $= 6 \times \$25$
$$= \$150$$

b) Total cost of magazines $= 3 \times \$15$
$$= \$45$$

(ii) Total cost of all books and magazines

$$= \$150 + \$45$$
$$= \$195$$

Discount $= 8\%$

∴ Total amt. Daniel actually paid

$$= \$195 - (8\% \text{ of } \$195)$$

$$= \$195 - (0.08 \times 195)$$

$$= \$195 - \$15.60$$

$$= \$179.40$$

6. (a)

$$x + 2y = 7 \quad - \quad ①$$
$$3x + y = 6 \quad - \quad ②$$

① × 3: $\quad 3x + 6y = 21 \quad - \quad ③$

② × 1: $\quad 3x + y = 6 \quad - \quad ②$

_____

③ − ②: $\quad 6y - y = 21 - 6$

$$5y = 15$$
$$y = \frac{15}{5}$$
$$y = 3$$

Substitute $y = 3$ into eqn ①

$$x + 2(3) = 7$$
$$x + 6 = 7$$
$$x = 7 - 6$$
$$x = 1$$

∴ $x = 1, \ y = 3$

(b) (i)a) Cost of pen = p

∴ Cost of calculator = p + 36

b) Total cost of pen and calculator

$$= p + (p + 36)$$

$$= p + p + 36$$

$$= 2p + 36$$

(ii) Total cost of pen and calculator = \$54

$$∴ \quad 2p + 36 = 54$$
$$2p = 54 - 36$$
$$= 18$$
$$∴ \quad p = \frac{18}{2}$$
$$p = 9$$

7 (a) (i)

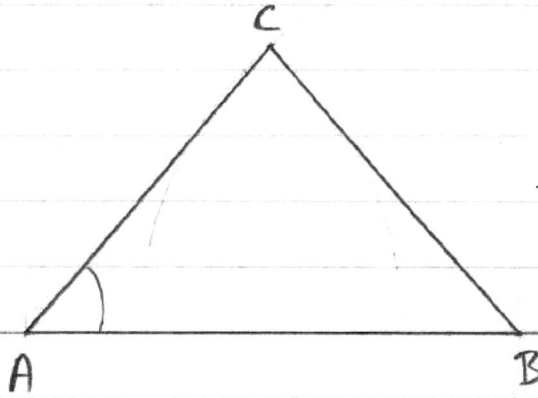

(ii) Angle BAC = 49.5°

(b) (i) $5^2 + 12^2 = XY^2$

$\Rightarrow \quad 25 + 144 = XY^2$

$\Rightarrow \quad 169 = XY^2$

$\Rightarrow \quad \sqrt{169} = XY$

$\Rightarrow \quad 13 = XY$

$\therefore \quad XY = 13cm$

(ii) $\dfrac{XZ}{13} = \sin 50$

$\Rightarrow \quad \dfrac{XZ}{13} = 0.7660$

$\Rightarrow \quad XZ = 13 \times 0.7660$

$\quad = 9.96cm$

8 (a) (i) Chris took 35 seconds to complete the race

(ii) After 20 seconds, Chris had run 125m.

$\therefore$ Distance from finish line $= 200 - 125$
$$= 75m$$

(iii) The two boys were the same distance from the starting point 12.5 seconds after the start of the race

(iv) During the race Chris fell. He stayed on the ground for 5 seconds (from $t = 10$ to $t = 15$)

(v) His average speed when he fell was 0 metres per second

(b) (i) Gradient of PQ = 2

(ii) Eqn of line PQ:

$$y - y_1 = m(x - x_1)$$

$$m = 2, \quad x_1 = 2, \quad y_1 = 7$$

$\therefore \quad y - 7 = 2(x - 2)$

$\Rightarrow \quad y - 7 = 2x - 4$

$\Rightarrow \quad y = 2x - 4 + 7$

$\Rightarrow \quad y = 2x + 3$

$\therefore$ Eqn of PQ: $\quad y = 2x + 3$

9. (a) $K(2,2)$; $K'(4,-4)$

(b) Column vector representing the translation:

$$\begin{pmatrix} 4 \\ -4 \end{pmatrix} - \begin{pmatrix} 2 \\ 2 \end{pmatrix} = \begin{pmatrix} 4-2 \\ -4-2 \end{pmatrix}$$

$$= \begin{pmatrix} 2 \\ -6 \end{pmatrix}$$

(d) Area of triangle KLM is 3 units.

Triangle K"L"M" is the image of triangle KLM after enlargement by a scale factor of 2.

The area of K"L"M" would therefore be 2^2 times the area of KLM.

$\therefore$ Area of K"L"M" $= 3 \times (2^2)$

$\quad\quad\quad\quad\quad\quad = 3 \times 4$

$\quad\quad\quad\quad\quad\quad = 12$ units

TEST CODE 01134020

FORM TP 2007103

MAY/JUNE 2007

CARIBBEAN EXAMINATIONS COUNCIL

SECONDARY EDUCATION CERTIFICATE
EXAMINATION

MATHEMATICS

Paper 02 — Basic Proficiency

Answer Sheet for Question 9 (c)   Candidate Number ....

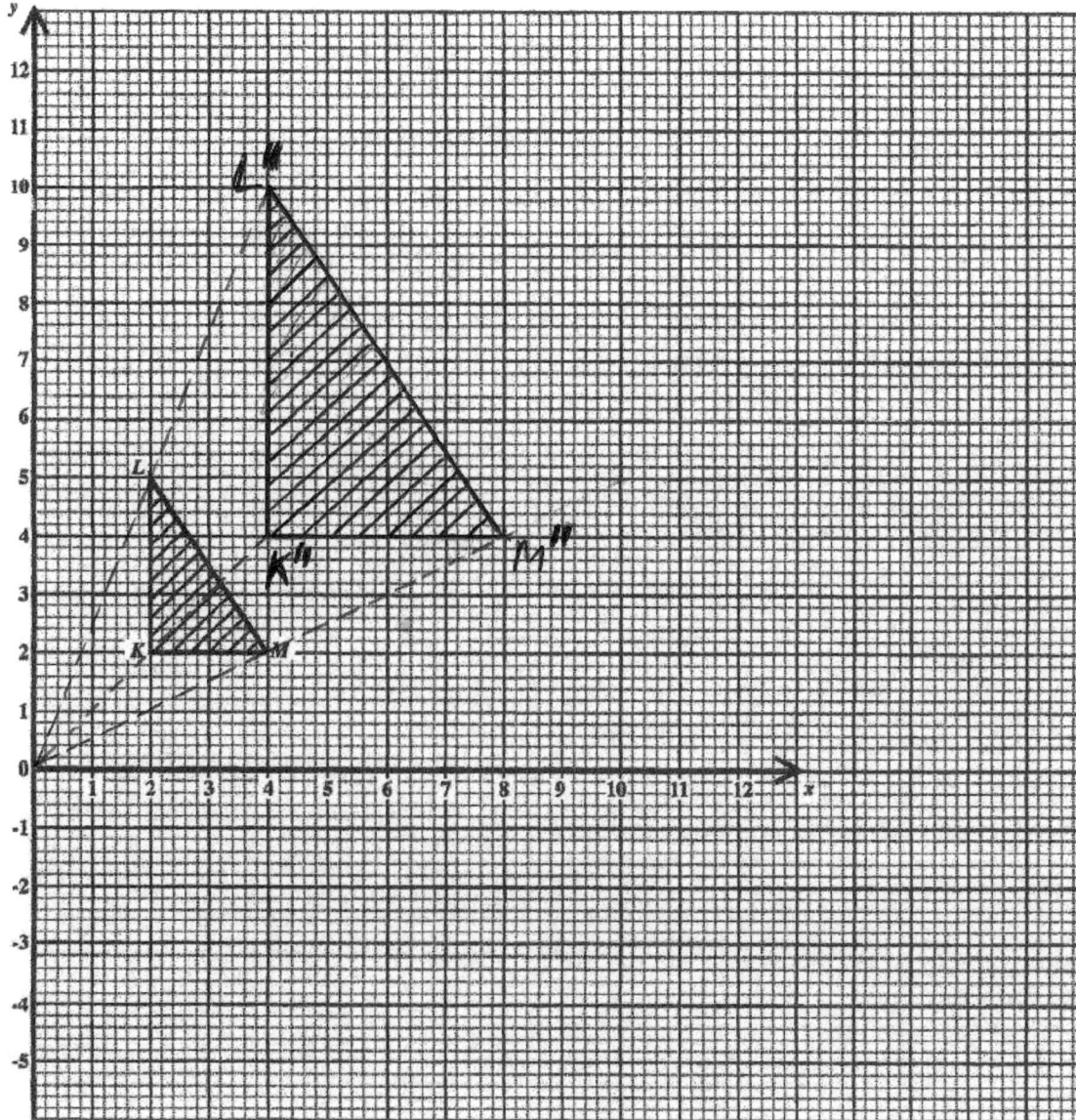

ATTACH THIS ANSWER SHEET TO YOUR ANSWER BOOKLET

10 (a) (i) No. of students whose favourite music is jazz:

$$= \frac{90}{360} \times 60$$

$$= \frac{{}^1\cancel{90}}{\cancel{360}} \times \frac{\cancel{60}^{15}}{1} = \frac{15}{1}$$

$$= 15 \text{ students}$$

(ii) Angle representing Reggae $= 360 - (162 + 90)$

$$= 360 - 252$$
$$= 108°$$

∴ Percentage of students whose favourite

Music is Reggae $= \frac{108}{360} \times \frac{100}{1}$

$$= \frac{\cancel{108}^3}{\cancel{360}} \times \frac{\cancel{100}^{10}}{1} = 30$$

$$= 30\%$$

(b) (i)

Height	Frequency
4	5
5	2
6	4
7	3
8	1
9	3
10	2

(ii) a) There are 20 plants, $\therefore$ Let $n = 20$

$\therefore$ The position of the median is given by

$$\frac{n+1}{2} = \frac{20+1}{2} = \frac{21}{2} = 10.5$$

$\therefore$ The median will be the average of the 10th and the 11th numbers.

The 10th number is 6, the 11th number is also 6

$\therefore$ The median height of the plants $= 6$ cm

b) P(plant is more than 8 cm tall)

$$= \frac{\text{No. of plants taller than 8cm}}{\text{Total no. of plants}}$$

$$= \frac{5}{20} = \frac{1}{4}$$

#

June 2008

1 (a) $\quad 1\frac{1}{3} - \left(\frac{5}{8} \div 1\frac{1}{4}\right)$

$\Rightarrow \frac{4}{3} - \left(\frac{5}{8} \div \frac{5}{4}\right)$

$\Rightarrow \frac{4}{3} - \left(\frac{5}{8} \times \frac{4}{5}\right)$

$\Rightarrow \frac{4}{3} - \left(\frac{\cancel{5}^1}{\cancel{8}_2} \times \frac{\cancel{4}^1}{\cancel{5}_1}\right)$

$\Rightarrow \frac{4}{3} - \frac{1}{2}$

$\Rightarrow \frac{(4)(2) - (3)(1)}{(3)(2)}$

$\Rightarrow \frac{8-3}{6}$

$= \frac{5}{6}$

(b) (i) Let no. of teachers = x

$$\therefore \quad x : 65 = 2 : 13$$

$$\Rightarrow \quad \frac{x}{2} = \frac{65}{13}$$

$$\Rightarrow \quad \frac{x}{2} = 5$$

$$\Rightarrow \quad x = 5 \times 2$$
$$= 10$$

$\therefore$ There were 10 teachers on Bus B.

(ii) Total no. of passengers = No. of teachers + No. of students

$$= (6 + 10 + 8) + (50 + 65 + 61)$$

$$= 24 + 176$$

$$= 200$$

(iii) % students $= \dfrac{\text{No. of students}}{\text{No. of passengers}} \times 100$

$= \dfrac{176}{200} \times 100$

$= 88\%$

2. (a) $a = 4$; $b = 5$; $c = -2$

    (i) $2b + c = 2(5) + (-2)$

$$= 10 + (-2)$$
$$= 10 - 2$$
$$= 8$$

    (ii) $\dfrac{c^2}{a} = \dfrac{(-2)^2}{4}$

$$= \dfrac{4}{4}$$
$$= 1$$

(b) $3(4y + 1) - 5y$

$\Rightarrow 12y + 3 - 5y$

$\Rightarrow 12y - 5y + 3$

$= 7y + 3$

(c) $\quad 2x + 15 \geqslant 7x$

$\Rightarrow \quad 15 \geqslant 7x - 2x$

$\Rightarrow \quad 15 \geqslant 5x$

$\Rightarrow \quad \dfrac{15}{5} \geqslant x$

$\Rightarrow \quad 3 \geqslant x$

$\Rightarrow \quad x \leqslant 3$

3. (a) (i) Amt. paid on hire purchase

$$= \$100 + (24 \times \$22.50)$$

$$= \$100 + \$540$$

$$= \$640$$

(ii) Extra amt. paid on hire purchase vs cash price:

$$= \$640 - \$510$$

$$= \$130$$

(b) Total amt. deposited after 6 months

$$= \$90 + (\$25 \times 6)$$

$$= \$90 + \$150$$

$$= \$240$$

(ii) Simple Interest $= \dfrac{\text{Principal} \times \text{Rate} \times \text{Time}}{100}$

$$= \dfrac{90 \times 4 \times \frac{1}{2}}{100}$$

$$= \dfrac{90 \times 4 \times 1}{100 \times 2}$$

$$= \dfrac{360}{200}$$

$$= \$1.80$$

∴ Amt. in bank after interest added

$$= \$240 + \$1.80$$

$$= \$241.80$$

4 (a)

$$x + 2y = 10 \quad - \text{①}$$
$$3x + y = 15 \quad - \text{②}$$

$\text{①} \times 1:$

$$x + 2y = 10 \quad - \text{①}$$

$\text{②} \times 2:$

$$6x + 2y = 30 \quad - \text{③}$$

$\text{①} - \text{③}:$

$$-5x = -20$$
$$x = \frac{-20}{-5}$$
$$= 4$$

Subst $x = 4$ into eqn ① :

$$4 + 2y = 10$$
$$\Rightarrow \quad 2y = 10 - 4$$
$$= 6$$
$$\Rightarrow \quad y = \frac{6}{2}$$

$$\therefore x = 4, \; y = 3$$

$$= 3$$

(b) (i) Sue's age $= p+3$

(ii) Joan's age $= 2(p+3)$

(iii) Sum of ages $= p + (p+3) + 2(p+3)$

$$= p + p + 3 + 2p + 6$$

$$= p + p + 2p + 3 + 6$$

$$= 4p + 9$$

(iv) Sum of ages $= 37$ years

$$\therefore \quad 4p + 9 = 37$$

$$\Rightarrow \quad 4p = 37 - 9$$

$$= 28$$

$$\therefore \quad p = \frac{28}{4}$$

$$= 7$$

$\therefore$ Dwight is 7 years old.

(9) (i) Total wage = Basic wage + Overtime wage

$\Rightarrow$ 285.50 = 248 + Overtime wage

$\Rightarrow$ 285.50 - 248 = Overtime wage

$\therefore$ Overtime wage = \$37.50

(ii) Overtime rate of pay = $\dfrac{\text{Overtime wage}}{\text{Overtime hours}}$

$$= \frac{37.50}{5}$$

$$= \$7.50/hr$$

(b) (i) Total value of US currency

$$= (2 \times 100) + (6 \times 50) + (4 \times 20) + (3 \times 10)$$

$$= 200 + 300 + 80 + 30$$

$$= \$610$$

(ii) US$610 = (610×2.70) EC dollars

$$= EC\$1,647$$

Bank charge = 2%
$$= 0.02 \times 1647$$
$$= \$32.94$$

∴ Value of EC currency received after deduction
of bank charges = $1,647 - $32.94

$$= EC\$1,614.06$$

6 (a) (i)

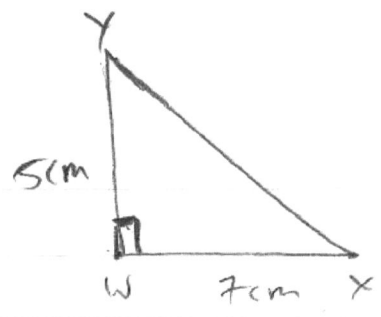

5cm

W   7cm   X

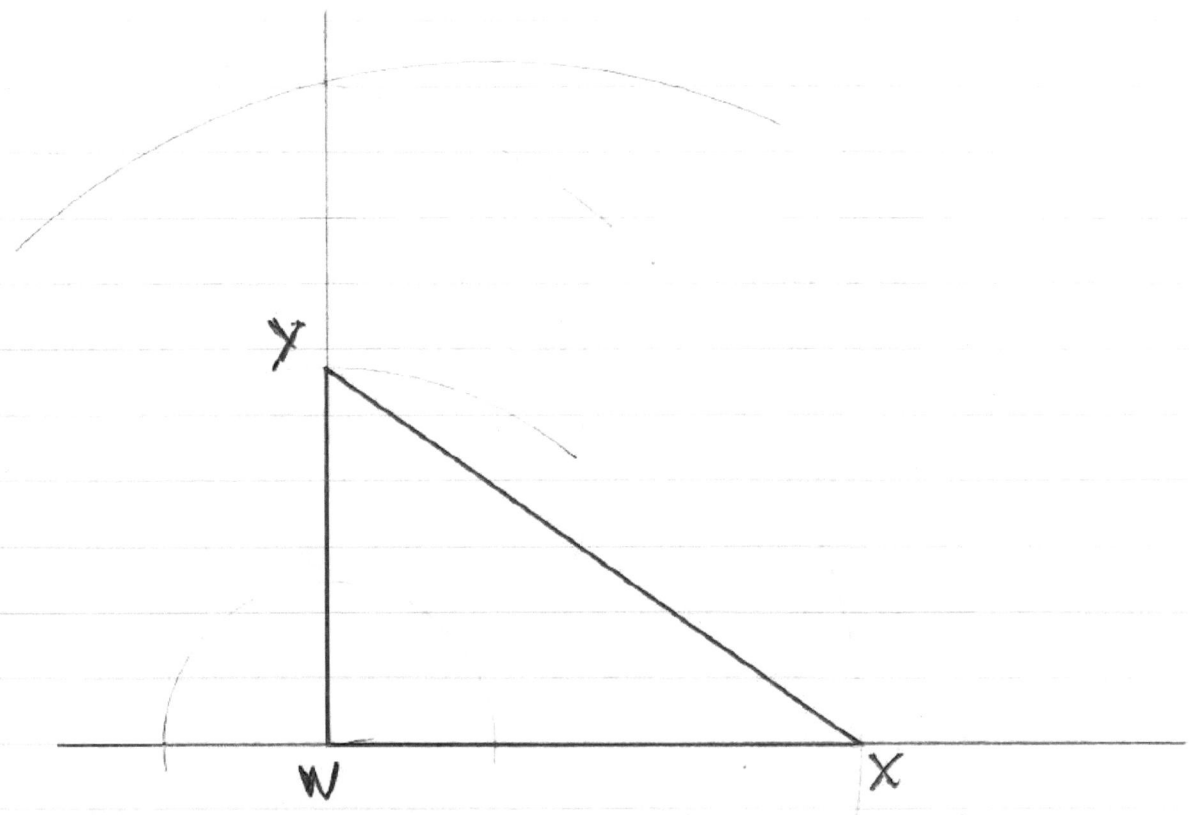

Y

W   X

(ii) XY = 8.7 cm

(b) (i) Let height ladder reaches up wall = h m

Using Pythagoras' Theorem :

$$h^2 + 2.4^2 = 3.6^2$$

$$\Rightarrow \quad h^2 = 3.6^2 - 2.4^2$$
$$= 12.96 - 5.76$$
$$= 7.2$$

$$\therefore \quad h = \sqrt{7.2}$$
$$= 2.7 \text{ m}$$

$$\therefore \quad h = 2.7 \text{ m}$$

(ii) $\quad \sin\alpha = \dfrac{2.4}{3.6}$

$$\therefore \quad \alpha = \sin^{-1}\left(\dfrac{2.4}{3.6}\right)$$

$$= 42^0$$

7 (a) $20km = (20 \times 100,000)cm$

$$= 2,000,000 \ cm$$

Distance on map $= \dfrac{2,000,000}{500,000}$

$$= 4cm$$

(b)(i) Area of triangle MJK $= \dfrac{4.5 \times 6.0}{2}$

$$= \dfrac{27}{2}$$

$$= 13.5 \ cm^2$$

(ii) Area of triangle KLM $= 26.26 - 13.5$

$$= 12.76 \ cm^2$$

(iii) $LM = 21.5 - (6.0 + 4.5 + 3.5)$

$\quad\quad\quad = 21.5 - 14$

$\quad\quad\quad = 7.5\ cm$

8 (a) (i) $P(6,-4)$; $Q(10,-4)$; $R(6,-2)$

(ii) PQR is mapped to P'Q'R by a translation of $\begin{pmatrix} -4 \\ 8 \end{pmatrix}$

(c) Area of $\triangle P'Q'R' = 4$ square centimetres

Area of $\triangle P''Q''R'' = 16$ square centimetres

TEST CODE **01134020**

FORM TP 2008098

MAY/JUNE 2008

CARIBBEAN EXAMINATIONS COUNCIL

SECONDARY EDUCATION CERTIFICATE
EXAMINATION

MATHEMATICS

Paper 02 – Basic Proficiency

Answer Sheet for Question 8

Candidate Number ....

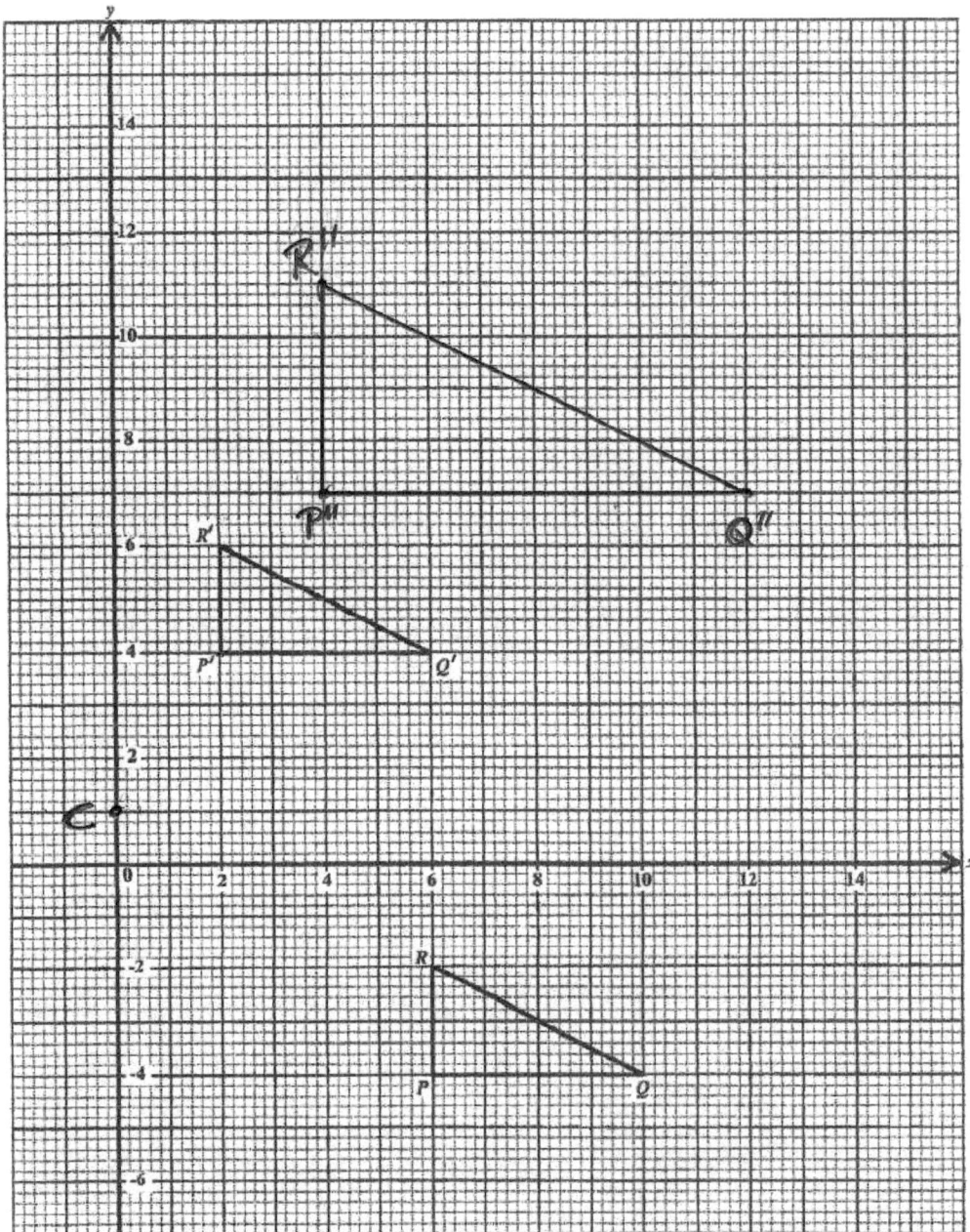

9. (a) (i) Gradient of $l = \dfrac{4-(-2)}{2-(-1)}$

$$= \dfrac{4+2}{2+1}$$

$$= \dfrac{6}{3}$$

$$= 2$$

(ii) Equation of l:

$$y-4 = 2(x-2)$$

$$\Rightarrow \quad y-4 = 2x-4$$

$$\Rightarrow \quad y = 2x-4+4$$

$$= \quad y = 2x$$

(b) (i)

x	-3	-2	-1	0	1	2
f(x)	-6	-1	2	3	2	-1

$f(x) = 3 - x^2$

$$f(-2) = 3 - (-2)^2$$
$$= 3 - 4$$
$$= -1$$

$$f(0) = 3 - (0)^2$$
$$= 3 - 0$$
$$= 3$$

$$f(1) = 3 - (1)^2$$
$$= 3 - 1$$
$$= 2$$

(iii) The line l and the graph $f(x) = 3 - x^2$ intersect at the points $(1, 2)$ and $(-3, -6)$

FORM TP 2008098

MAY/JUNE 2008

CARIBBEAN EXAMINATIONS COUNCIL

SECONDARY EDUCATION CERTIFICATE
EXAMINATION

MATHEMATICS

Paper 02 – Basic Proficiency

Answer Sheet for Question 9   Candidate Number ....

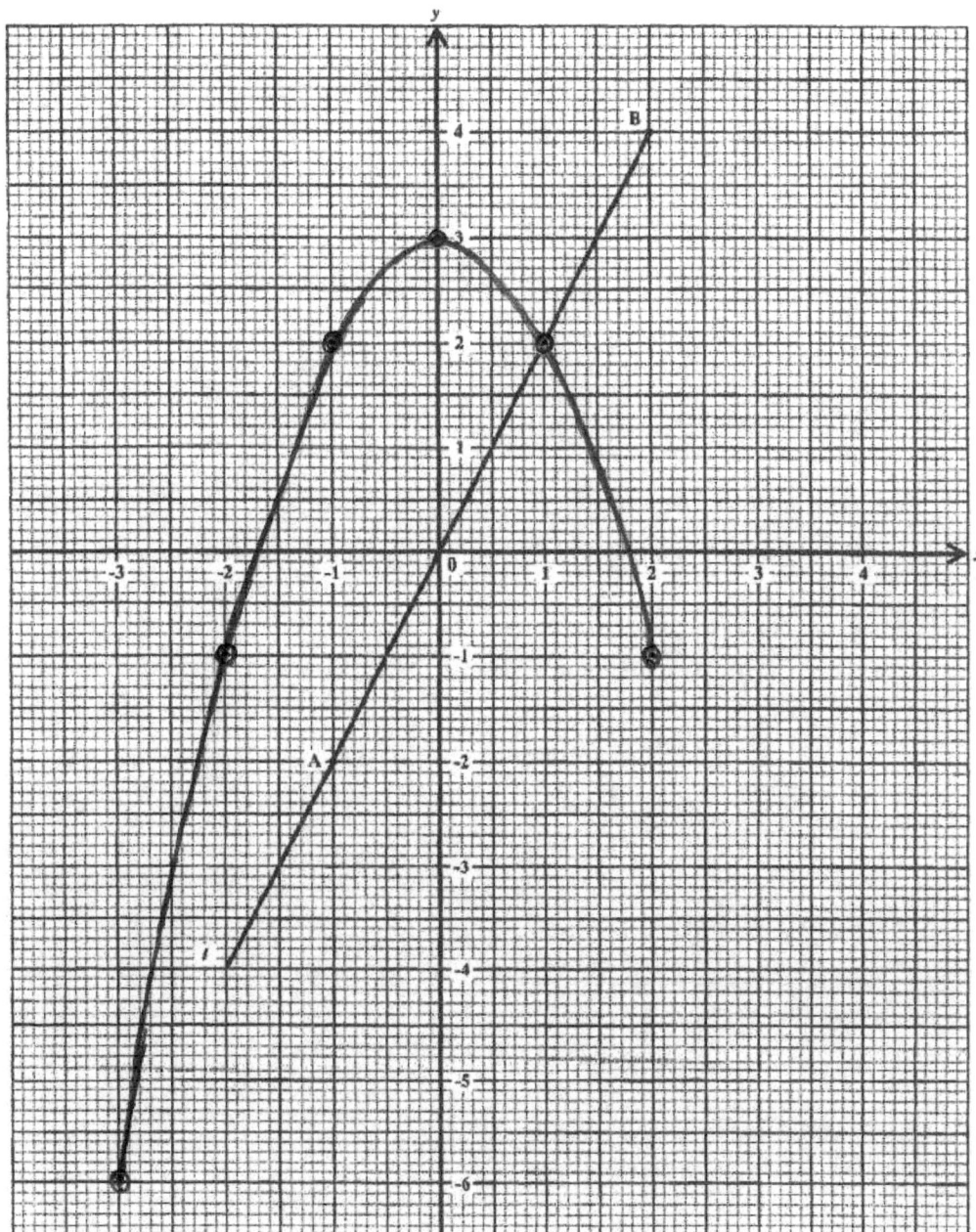

ATTACH THIS ANSWER SHEET TO YOUR ANSWER BOOKLET

10 (a) (i) Total number of runs = 55×3
$$= 165 \text{ runs}$$

(ii) Total number of runs after the fourth
match = 61×4
$$= 244 \text{ runs}$$

∴ No. of runs scored in the fourth
match = 244 − 165

$$= 79 \text{ runs}$$

(b) (i) No. of children who play soccer

$$= \frac{150}{360} \times 120 \qquad \frac{\overset{50}{\cancel{150}}}{\underset{3}{\cancel{360}}} \times \frac{\overset{1}{\cancel{120}}}{1}$$

$$= 50 \text{ children}$$

(ii) $\qquad \dfrac{x}{360} \times 120 = 46$

$\Rightarrow \qquad \dfrac{x}{360_3} \times \dfrac{\cancel{120}^1}{\cancel{}} = 46$

$\Rightarrow \qquad \dfrac{x}{3} = 46$

$\Rightarrow \qquad x = 3 \times 46$
$\qquad\qquad = 138$

$\qquad \therefore \quad x = 138^{\circ}$

(iii) No. of children who dance $= 120 - (50 + 46)$
$\qquad\qquad\qquad\qquad\qquad\qquad = 120 - 96$
$\qquad\qquad\qquad\qquad\qquad\qquad = 24$

$\therefore$ P (child dances) $= \dfrac{\text{No. of children who dance}}{\text{Total no. of children}}$

$\qquad\qquad\qquad\qquad = \dfrac{24}{120} \qquad = \dfrac{1}{5}$